GUIDE TO OBSERVING THE MOON

British Astronomical Association

ENSLOW PUBLISHERS, INC.

Bloy St. & Ramsey Ave.
Box 777
Hillside, N.J. 07205
U.S.A.

P.O. Box 38
Aldershot
Hants GU12 6BP
U.K.

Library of Congress Cataloging in Publication Data

British Astronomical Association
 Guide to observing the moon.

 Includes index.
 1. Moon--Observers' manuals. I. British Astronomical
Association. Lunar Section.
QB581.G83   1986   523.3'028   83-1610
ISBN 0-89490-085-4

Printed in the United States of America

10 9 8 7 6 5 4 3 2 1

CONTENTS

FOREWORD

Telescopic studies of the Moon began in the early 17th century. Today, men have landed on the lunar surface, and the manned and unmanned space probes have provided highly accurate maps of the lunar surface; but the Earth-based observer still has work to do, and in this field the Lunar Section of the British Astronomical Association has played a notable role. This book will, we hope, provide a useful guide for all those who are interested in the practical observation of our satellite.

Patrick Moore
Past-President
British Astronomical Association

1

INTRODUCTION
by G.W. Amery

The Lunar Section of the B.A.A. was formed in 1891, shortly after the formation of the Association itself. It has been an active observational Section ever since. Under its first directors, T. Gwyn Elger (1891-1896) and Walter Goodacre (1896-1938), it was largely concerned with cartography and both these directors published lunar maps which were regarded as standard works. At this time the scope for the amateur was unlimited.

During the war years 1939-1945 activity in the Section was understandably at a low ebb. After the war, however, H.P. Wilkins became director and, with the help of other active observers such as Patrick Moore, D.W.G. Arthur, E.A. Whitaker, and F.H. Thornton, the Section was rebuilt. Charting of the Moon continued and the Section published its own journal *The Moon* which carried many drawings made by its observers. Observations of anomalous appearances reported by the cartographers from time to time led to the realization that the Moon might not be completely inert, and this initiated the field of transient lunar phenomena (TLP) studies (usually LTP–lunar transient phenomena–in the United States), now so much a part of the modern Lunar Section's work.

Then came the space age. The Moon was naturally the prime target; it was duly photographed in detail from close quarters, and finally the Apollo missions enabled samples and other valuable data to be obtained. After this period many people questioned the value of further amateur observation of the Moon. Thanks largely to directors such as Patrick Moore and R.C. Maddison, these doubts were shown to be unfounded. All that was needed was a change of emphasis. It was recognized that cartographic work in the original sense was now obsolete, so the Section concentrated on observing time-dependent phenomena. The timing of lunar occultations became an important part of the programme, and the TLP Network was formed, whereby observers could be telephoned when confirmatory observations were required at short notice.

During recent years spacecraft have opened up the Solar System to an extent never dreamed of in earlier years. Not only the Moon but Mercury, Venus, Mars, Jupiter, and Saturn have yielded secrets forever hidden from Earth-based telescopes. Yet these probes cannot provide answers to every question. The need for patient, vigilant observation on the part of Earth-based observers still remains, and, especially where the Moon is concerned, much of this lies within the capability of amateurs.

The observational programmes of the post-Apollo Lunar Section have been organized to enable them to participate in serious research projects or, if they prefer, to enjoy observing the Moon simply for pleasure. Much of the scientifically useful work still concerns the observation of time-dependent phenomena such as TLPs or occultations.

The old cartographic programmes, once the lifeblood of the Section and which contributed so much to our knowledge of the Moon, are now obsolete, having been superseded by the highly successful Lunar Orbiter photography. However, this does not mean there is no further use for amateurs who enjoy drawing and photographing lunar surface features. Firstly, it is as true today as it ever was, that there is no better way of learning about the Moon than by drawing it. Even if the observer's artistic skills are limited, an attempt to record what is seen impresses detail on the mind as nothing else can. Secondly, those who enjoy sketching (and many can acquire the necessary skill through practice) can now contribute to an organized programme whereby their work is arranged in lunation sequences to record the changing appearance of a feature as seen from Earth. It is believed that long-term studies of this kind, supported by a full survey of the relevant Lunar Orbiter photography and earlier Earth-based observations, can still contribute to our knowledge.

Monitoring of the lunar surface for transient phenomena is a major project now carried out on a routine basis. An observer Network is organized to provide confirmation of suspected TLP and all confirmed reports are retained by the Section for future analysis. Attempts are now being made to quantify albedo measurements by the use of the crater extinction device. The existence of recurrent transient phenomena is now recognized and their locations charted.

The timing of lunar occultations also forms a major Section project. Although the Earth-Moon distance can now be determined with great accuracy by laser ranging techniques, there is still a need to relate these measures to the fundamental system. Occultation timings can be used for this purpose and also to derive ΔT, the difference of ephemeris time and Universal Time. In particular, the timing of immersion/emersion sequences occurring during a grazing occultation requires the use of portable telescopes and is ideally suited to amateurs, especially where groups can be organized for the purpose. The data acquired can be most useful if they form part of an international input. For this reason observations are regularly despatched to the Japanese Hydrographic Department in Tokyo for reduction. The Section now acts as the reception centre for UK observations and computes preliminary reductions for its observers.

A growing collection of books and archival material is in the care of a Librarian and can be made available to members at his discretion. Assistance with technical and photographic queries is available and meetings are held regularly in London or at the invitation of provincial societies.

Such then is the Lunar Section of the post-Apollo era. It is perhaps unfortunate that, just when the space programmes were getting under way, serious problems have imposed restraints on further missions to the Moon. Fortunately the amateur is not beset by such problems; only clouds can prevent his engaging in the observational programmes which lie within his scope.

We also believe there is a need for a general handbook both to help newcomers to lunar work and to summarize our current observational programmes and techniques. The first edition of the Section's handbook was published in 1972 as *Guide for Observers of the Moon.* It was revised and updated in second and third editions. Now, in collaboration with Enslow Publishers, we are publishing this expanded and re-set version under the title *Guide to Observing the Moon.* Most of the chapters have been written by dedicated amateurs, all specialists in their own field. We have also been fortunate to be able to include chapters by two eminent professionals. We hope the book will be found useful by other amateurs, especially those entering the fascinating world of lunar studies for the first time.

I would like to express gratitude and appreciation to all who have made this book possible and a hope that it will provide a stimulus to all those who share our interest in the Moon, whether in the United Kingdom, the United States, or elsewhere.

Geoffrey W. Amery
Director, British Astronomical
Association Lunar Section

ABOUT THE AUTHOR

Although he has had a life-long interest in astronomy, Geoff Amery began a serious study of the subject in later life, progressing from the use of a small refractor to building his own 25 cm reflector. An industrial chemist by profession and member of the B.A.A. since 1969, he joined the Lunar Section in 1971 and specialised in the observation of lunar occultations. From 1974-1978 he co-ordinated the newly formed occultation sub-section within the Lunar Section and worked to promote this study as an amateur project in co-operation with professional astronomers. He was appointed Director of the Lunar Section in 1978 and continues to carry out occultation and TLP studies.

2

EQUIPMENT
by C.J. Watkis

Before anyone can observe the features of the Moon, some kind of instrument is necessary. The Moon is the nearest celestial object in the sky, a mere 384,400 km (about a quarter of a million miles) away, yet a reasonable degree of magnification is required in order that mountains and craters can be seen. A pair of binoculars can be used for general views, but the usual type owned by most people (10 X 50 or 8 X 40) will show a disappointingly small image. More powerful types are required, e.g., 15 X 40 or 20 X 60, which are very expensive. Therefore, if one contemplates buying new equipment, a telescope should be considered as it would be a better buy for lunar observations.

CHOICE OF INSTRUMENT
For a beginner the first criterion for observing the Moon is to get to know the lunar surface with its myriad of mountains and craters. To do this a small telescope which will provide sufficient magnification and clarity will be required.

The smallest telescope to be considered would be a 50 or 75 mm (2- or 3-inch) refractor or a 100 mm (4-inch) reflector giving usable magnifications between 40 to 80 times. This will be sufficient to show all the major features. Considerable care should be taken before making any purchase as it is only too easy to buy the proverbial 'pig in a poke'. There are many cheap telescopes on the market, and it is always prudent to get expert advice if possible. If this is not possible, beware of judging the instrument only by its outward appearance, for it is the quality of the optics that makes an instrument good or useless. One particular point to watch when selecting a refractor is whether a circular diaphragm has been inserted immediately behind the object glass, as this reduces the effective diameter of the lens and therefore reduces its light gathering ability. It also indicates that the object lens is of poor quality since it is designed to minimize a multitude of aberrations which would make the image impossible to view clearly. Also, beware of the 100 mm (4-inch) reflector, for unless the mirror is of reasonable quality, this will also be a waste of money. It follows that you get what you pay for and, unfortunately, there is no cheap way of obtaining a good optical instrument.

One can cut cost by grinding one's own mirror, which is not difficult but very time-consuming, or one can purchase a finished mirror to build into a home-built tube and mounting. If observations are to be made which are to be of any use in connection with one of the Sub-Section programmes, the largest telescope that one can afford should be aimed at. This should be at least a 150 mm (6-inch) reflector, but preferably a telescope with a 200 mm to 300 mm (8- to 12-inch) mirror.

EYEPIECES

It is commonly thought that a telescope is only of use if it is capable of high magnification. However, when observing the Moon, although a reasonable degree of magnification is needed, the amount is dependent on the size of the telescope and the 'seeing' conditions. The size of the object lens or mirror will dictate the resolving power of the instrument, i.e., the ability to distinguish between two points close together. Similarly, the turbulent atmosphere influences the seeing conditions, which can make the Moon's image crisp and steady or wobbly like a jelly.

Therefore, a selection of eyepieces to cover all eventualities is required, but do not be tempted to collect a battery of these, for good quality ones are expensive, and not all of them will be used. The most useful are a 25 mm (1-inch) and a 12.5 mm (½-inch), used in conjunction with an achromatic 3X Barlow lens. Thus almost any telescope will have a range extending from the minimum useful magnification up to almost the maximum the instrument will be able to handle.

MOUNTINGS

Irrespective of whether one is using a pair of binoculars or a telescope for observations, a rigid support is essential. The support must be rigid to prevent the image from dancing all over the field of view with every puff of wind, yet so constructed that the telescope can be moved to any part of the sky and held there. This is of great importance when high powers are used, for the field of view decreases as the image size increases.

Mountings are of two basic forms, (i) the *altazimuth,* which enables the instrument to be moved parallel to the horizon and vertically at right angles to it, and (ii) the *equatorial,* where the vertical axis is inclined so that it is parallel to the Earth's axis. The latter form enables a celestial object to be followed from rising to setting by turning only one axis, whereas an altazimuth requires constant adjustment in both motions to do the same job.

For small telescopes, such as a 50 mm or 75 mm (2- or 3-inch) refractor or 100 mm (4-inch) reflector, an altazimuth mounting can be used, but for anything larger, an equatorial mounting is almost a necessity.

Dobsonian Mounting

The Dobsonian mounting is a modified form of the altazimuth mounting. It was developed by an American amateur astronomer. Its main assets are its strength, rigidity, portability, and compactness, all features that are beneficial to those who live in well-lit urban areas and need to travel to dark, country observing sites.

Its fabrication is well within the capabilities of the handyman and it may be constructed from chipboard or sheet plywood, together with a few bolts and pieces of Teflon for bearing surfaces.

It is designed specifically for use with short-focal-length reflectors, which can have mirrors of any size from 150 mm (6 inches) to 450 mm (18 inches). Of course being of the altazimuth principle, photography of anything except bright objects, such as the Moon, is not possible for the reasons already stated.

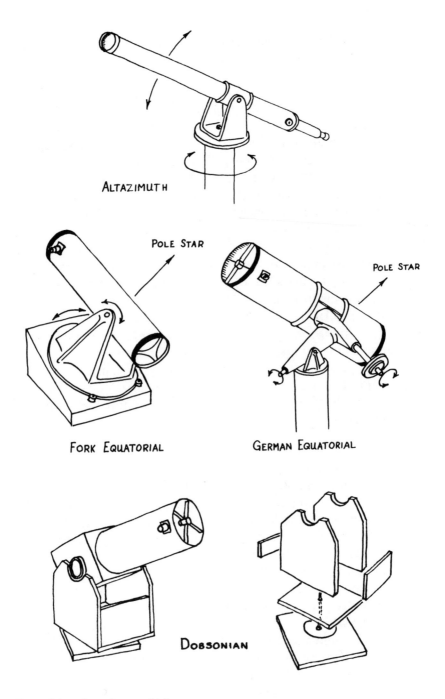

ALTAZIMUTH

POLE STAR

FORK EQUATORIAL

POLE STAR

GERMAN EQUATORIAL

DOBSONIAN

Figure 2-1. Several types of telescopes.

REFERENCE MATERIAL

As well as notebooks and sketch pads, a lunar map is essential. There are many available on the commercial market at a reasonable cost. Bert Chapman's 'Map of the Moon' (given here in sections as an appendix) is very good, showing a large amount of detail without unnecessary overcrowding, and has been specially drawn to be of use to beginners.

Books regarding the Moon are expensive, but some should be found on the shelves of any astronomical society library. Those in the Recommended Further Reading of this book are suggested. The older books may be out of print but, again, try your library.

Finally, remember that all astronomical observations normally take place in the dark, frequently when it is damp and cold. These conditions persist irrespective of whether you are an amateur with a 75 mm (3-inch) refractor in a back garden or a professional with a 2.5 metre (100-inch) telescope on a mountaintop, so wrap up warm before you observe. This is essential since, unless you are comfortable, you will not be able to concentrate on making accurate observations, sometimes necessarily for prolonged periods.

ABOUT THE AUTHOR

A member of the B.A.A. Instruments and Observing methods committee, Chris Watkis' special interest is in building his own equipment. He has built his own telescopes and mounts and more recently has undertaken the construction of electronic equipment. Until recently he was Curator of Instruments for the B.A.A.

3

EXPLORING THE MOON
by K.W. Abineri and C.J. Watkis

Of all the celestial bodies, the Moon presents by far the most detailed surface for visual study. This is due to its relative proximity to the Earth, the variable appearance of its terrain under different angles of illumination, and the lack of any detectable atmosphere to mar the clarity of its features. Although, superficially, the numerous 'craters' appear to be stereotype structures, closer inspection shows remarkable variation in form. In addition, there are many other types of features to be seen on the lunar surface, including the extensive dark plains (the maria), mountain ranges, valleys, rilles, faults, hillocks, domes—and the 'rays', dark areas, bright areas, streaks, and spots that are so conspicuous at Full Moon. Not only is the Moon regular in its cycle of phases, but is suitably placed for observation on frequent occasions (weather permitting!). Much of the surface detail can be seen with quite a small telescope; indeed a good quality 150 mm (6-inch) reflector, when used with good 'seeing', gives a fine view of the lunar landscape, presenting a picture which varies throughout the lunation due to the changing conditions of illumination and which is too complex to be known completely by any one individual observer. In the United States lunar observers have not agreed to the use of any single scale to define seeing conditions, but in the United Kingdom a scale originally devised by the Greek-French astronomer Eugene Antoniadi is used with conspicuous success.

The Antoniadi Scale

Numeral   *Seeing Conditions*

1.   Perfect conditions without a quiver in the atmosphere.
2.   Slight atmospheric undulations, with moments of calm
      lasting several seconds.
3.   Moderate seeing, with larger air tremors.
4.   Poor seeing, with constant undulations.
5.   Very bad seeing, scarcely allowing the making of a rough
      sketch.

THE INITIAL PROGRAMME

When you are starting on the first stage of lunar exploration, it is
most interesting to follow the shadow terminator across the Moon
during successive lunations. In this way you will become familiar
with the main topographical features and learn to identify them on
the map. When this has been done for several lunations, you will
begin to note differences in appearance due to varying libration,
especially in the limb regions. Later on, attention should be given
to features that are away from the shadow terminator and can be
discerned only by the differences in the intensity of the light
reflected by the lunar surface. At the Full phase, shadows are not
seen on the lunar disk (except in some areas very close to the limb),
since in the central regions the angle of illumination is far too high,
and away from the centre the features, as seen from the Earth,
obscure their own shadows. Under these conditions it is often very
difficult to identify some of the objects which are prominent when
close to the shadow terminator.

In the early part of the lunation, i.e., shortly after New Moon,
formations on the eastern limb will be seen under sunrise illumination. Search for the central mountain and rille in Petavius; these
are quite easy even in a small telescope. The rilles and other floor
details in Furnerius will require very favourable conditions for their
observation. Vendelinus and Langrenus contain also fine detail on
their floors and in their interior slopes. The boundary walls and
valleys of the Mare Crisium are complex and interesting to study at
sunrise. Note also any signs of ridges or ghost craters on the mare
surface.

On successive evenings, as the sunrise terminator moves westward, study the great damaged crater Janssen with its rilles and satellite craters, the extraordinary Rheita Valley with its neighbouring valleys and craters, the complex crater Reichenbach, Goclenius and its rilles, the Mare Fecunditatis, including Messier and its strange companion-crater, Proclus and its surroundings, as well as Hercules with its interior detail. Follow the shadow terminator into the polar regions. If libration is favourable for the south pole try to identify craters Boussingault, Boguslawsky, Demonax, and Schomberger. Note the illumination of isolated mountain peaks in the so-called Leibnitz Mountains. If libration is suitable for the north pole there are many interesting formations to be seen, including Meton, Scoresby, Main, and others.

Features nearer the centre of the visible disk are thrown into relief as the lunation proceeds. Examine the Mare Tranquillitatis, the Mare Serenitatis, the prominent craters Theophilus, Posidonius, and Aristoteles. The Altai Scarp, Piccolomini, Fracastorius, and the Mare Nectaris are very impressive at sunrise.

Maurolycus, Stöfler, Licetus, and Walter, in the central-southern highlands, when close to the sunrise terminator, all show structural features of interest, even when viewed through a small telescope. This applies equally to Albategnius, Alphonsus, and the great enclosure Ptolemaeus in the central regions, as well as to numerous other formations further north. Identify the Hyginus and Triesnecker rilles if seeing is good; explore the wonderful lunar Apennines and Alps on the borders of the Mare Imbrium. Near the First Quarter phase in the spring months you will see all the above features to the best advantage. The Alpine Valley and the very dark-floored crater Plato deserve special attention.

As the terminator moves further westward, watch the sunrise on Moretus, Clavius, Maginus, Tycho, Copernicus, and the Sinus Iridum. To observe the early sunrise in Clavius is certainly one of the most impressive sights, especially in a telescope of moderate aperture, since the western walls catch the sunlight to form a delicate and complex ring of illuminated peaks extending far into the unilluminated portion of the disk. Also at this time, the largest ring on the central floor forms a delicate circle of light on a small portion of the faintly illuminated interior. As the sunrise proceeds, the western illuminated ramparts become broader and more conspicuous and

illumination at the centre spreads westward, until the western floor
is free from continuous shadow. The details on the central and
western portion of the floor can be seen by virtue of their individual
shadows, and the shadows of the eastern walls form a very complex
boundary across the eastern floor. Even at the present time, a
precise and detailed study of this phenomenon by an experienced
observer might improve our knowledge of the topography of Clavius.
Maginus is equally spectacular at sunrise.

Similar observations can be made on Schiller, the Mare Humorum,
with its neighbouring formations Gassendi and Mersenius, or on
Kepler and Aristarchus. The latter is a particularly interesting lunar
feature renowned for its 'bands' and Transient Phenomena. As the
terminator approaches the western limb, some time before Full, try
to observe the sunrise on Schickard and Grimaldi; these splendid
enclosures have many fine interior details, which are interesting test
objects for the novice. If libration is favourable for the western (IAU)
limb, search for the very large crater Bailly; however some experience
is required before the new observer concentrates on the limb areas,
since varying extents of libration give rise to many problems in
identification and interpretation.

Having followed the sunrise shadow terminator across the Moon
during the first half of the lunation, note the changed appearance of
features at Full Moon. Formations which were prominent under
low-angle illumination are difficult to identify; for example, Maginus
is hidden by the conspicuous Tycho ray system. Indeed the ray
systems of Tycho, Copernicus, Aristarchus, Kepler, Proclus, etc.,
now dominate the lunar landscape, together with numerous bright
craters, white spots, streaks, dark areas, 'banded' craters, etc. The
range of intensities at Full Moon is far better appreciated by looking
at the Moon through a telescope than by studying photographs, since
the human eye is very capable of adapting to this type of observation.

After Full Moon, the sunset terminator again throws into relief
those magnificent east-limb formations including Furnerius, Petavius,
Langrenus, the Mare Crisium, etc. This opposite illumination
produces a contrasting aspect; however, if the season is favourable
(e.g. in the autumn), and libration is satisfactory, the most impressive
views of these regions will be obtained, especially after midnight with
the Moon high in the clear sky. If libration is near maximum, this
is the best opportunity to study the limb formations, such as the

Mare Australe, Wilhelm Humboldt, Hecataeus, Kästner, the Mare Smythii, Gauss, and perhaps the interesting Mare Humboldtianum further north. Again, if libration is suitable, explore the polar regions under the opposite angle of illumination. You may find some difficulty in identifying Demonax in the south polar regions shortly after Full Moon, but it is a valuable exercise to attempt this, as it demonstrates the problems encountered when studying the areas most affected by libration.

Janssen at sunset deserves special attention, since you will now have a much better view of its remarkable rille system. Perhaps sunset illumination on Maurolycus is one of the finest of all the lunar scenes for the visual observer with a moderate-aperture telescope. The western walls of this magnificent crater cast very complex shadow forms over the western floor and the curious southern extension. With the Moon high in the sky in the autumn, this is a sight not to be missed, as the decreasing angle of the lighting accentuates fine detail in the central mountains, in the eastern floor, and in the eastern interior slopes. Also Maginus at Last Quarter is an object worthy of the enthusiast who is prepared to observe into the early hours on a cold, clear November night. The floor and wall detail supply excellent test objects.

In general so much is to be seen by the new observer as he follows the terminators patiently on every suitable occasion, that it is not possible to do full justice here to the scope of this initial programme. In addition, if the programme is extended to include the study of features away from the shadow terminators, a good understanding will be acquired of the effects of varying conditions of illumination on the appearance of lunar surface detail. This experience should be of great value to the novice in preparation for more systematic observations later in the various Lunar Section programmes. These programmes include the search for Transient Lunar Phenomena (TLPs) and the study of slow variations in appearance due to changes in illumination.

During this initial programme the new observer should make good use of the small-scale map and the useful data in the Lunar Section Circulars, which give information on the phases of the Moon, libration and the selenographic longitude of the morning terminator. The B.A.A. *Handbook* should also be consulted.

RECORDS

No matter what observations are made, they are of no use to anyone unless records are kept. Details of time and date are obvious requirements to pinpoint when an observation is made, but other details are necessary to give a fuller description. These will vary according to the method of observation adopted.

The methods most commonly used are written comments, drawings and photographs. The written comments should include

1. Type and size of instrument
2. Type and focal length of the eyepiece used
3. Magnification
4. Time in Universal Time, which should be checked
   against the telephone time or radio pips
5. Area of the Moon under observation

It should always be remembered that a picture can 'say' much more than the equivalent space in words and that a drawing or a photograph is therefore always an enhancement. If a drawing is to be made, use a reasonable scale, and do not be tempted to draw too large an area. Choose a single object and include as much detail as you can see.

When taking photographs, additional information is always useful and the relevant notes should be made at the time the picture is taken. Never trust to memory as you will not be sure of the details even if only five photographs have been taken.

Suggested information should include:

1. Type of film—i.e., colour or black-and-white, its speed
   (ASA)
2. Exposure in seconds or fractions of a minute
3. Equipment (camera, with or without lens; eyepiece and
   Barlow lens projection)
4. Magnification, or kilometres (or miles) per millimetre (or
   inch). (Calculation will be necessary if projection is used.)

If you do your own developing and printing, the following points should be included:

5. Type of developer used
6. Exposure for the print
7. Type and grade of the paper used
8. Magnification of the enlargement
9. Developer temperature

All this information will enable you to evaluate the results at a future time. Since a great deal of information has to be collated at the time the observation is made, a preprinted form is a good idea.

When any observations have been made, do not carelessly put them in a drawer but keep them in a file or exercise book, in date order, where they can be referred to at a future time. Who knows, they may be of great significance!

ABOUT THE AUTHORS

An industrial chemist by profession, Keith Abineri has been a member of the B.A.A. for many years. He has made many hundreds of lunar observations using a 200 mm reflector and was especially active during the period 1947-1960. His special interest is in the lunar limb regions and other selected features. Other interests include life and earth sciences, music, gardening and mountain climbing.

A member of the B.A.A. Instruments and Observing methods committee, Chris Watkis' special interest is in building his own equipment. He has built his own telescopes and mounts and more recently has undertaken the construction of electronic equipment. Until recently he was Curator of Instruments for the B.A.A.

4

DIRECTIONS AND CO-ORDINATES
by H.R. Hatfield

EAST AND WEST ON THE MOON

The International Astronomical Union has recommended that on the Moon, east should be in the direction in which the Sun rises, and west in the direction in which the Sun sets. Thus, Mare Crisium is in the east and Mare Humorum and the large crater Grimaldi are to the west. The Sun illuminates the eastern (Mare Crisium) limb of the Moon first at the start of each lunation. Before 1961, east and west on the Moon were directions as seen from the Earth; using this, the classical system, Mare Crisium was near the western limb and Grimaldi was near the eastern limb.

To avoid confusion, observers should use the present system and refer to 'east (IAU)' or 'west (IAU)'. If for any reason it is found necessary to use the classical system, the fact must be clearly stated. In this respect, new observers are advised to read the section below on libration with great care.

NORTH AND SOUTH ON THE MOON

It is fortunate that north and south on the Moon are the same for the Earthbound observer and for the astronaut on the Moon. Some modern lunar maps show north at the top, and have to be turned upside down to conform to what is seen in most telescopes (for observers in the Northern Hemisphere).

LATITUDE AND LONGITUDE ON THE MOON

A point near the crater Bruce is the origin of lunar co-ordinates. This point is at selenographic latitude 0° and selenographic longitude 0°. Northerly latitudes are in the direction of the north pole and the crater Plato. In the *Astronomical Almanac* and the B.A.A. *Handbook*, westerly longitudes are classical and refer to the direction of the Mare Crisium. On many modern maps this convention has been reversed in accordance with the IAU ruling. Note that this point at latitude 0°, longitude 0° will be in the centre of the Moon's disk only if the libration happens to be zero.

LIBRATION, OR THE EARTH'S SELENOGRAPHIC LONGITUDE AND LATITUDE

Libration is an apparent rocking motion imparted to the Moon as seen from the Earth. It may be split into two components, libration in latitude and libration in longitude (as in the *Astronomical Almanac,* the B.A.A. *Handbook* and the B.A.A. Lunar Section *Circulars*). Libration in latitude is caused by the fact that the Moon's axis of rotation is not at right angles to its orbital plane. Libration in longitude is caused by the fact that the Moon's speed of rotation about its axis is virtually constant, while its orbital velocity varies (being fastest at perigee and slowest at apogee). The total libration (i.e., the vector sum of libration in latitude and longitude) can be anything up to about 10°; 7° or 8° is considered to be quite large.

When the libration in latitude is positive, the north pole of the Moon is tilted toward the observer. When the libration in longitude is positive, the western (classical) limb (i.e., eastern [IAU] limb) is tilted toward the observer, and the Mare Crisium will be well presented. Note that both the *Astronomical Almanac* and the B.A.A. *Handbook* refer to east and west on the Moon in the classical sense so far as libration is concerned.

The direction and amount of the libration can best be determined from the tabulated values by drawing a graph, which may also give additional information, such as the Moon's age and the times of perigee and apogee. To make your graph proceed as follows:

    1. Draw N-S and E-W axes, intersecting in the middle of the sheet; you will need seven units each way in the N-S direction and 8 or 9 each way in the E-W direction; these points mark degrees.

2. Label the axes: + Lat., 'North limb exposed'; − Lat., 'South limb exposed'; + Long., 'West limb exposed'; and − Long., 'East limb exposed'; as shown in the accompanying diagram. Make a firm mental note that the Mare Crisium is near the west limb (classical).

3. Plot values of libration, obtained from the *Circular* or the *Astronomical Almanac*, against time. For most purposes it will probably be enough to plot every three days or so.

4. Join the plotted points. Then the direction of libration at any time will be the direction of the relevant point on the curve as seen from the intersection of the graph axes, and the amount of libration will be the distance of this point from the intersection of the axes, measured in the units used for plotting.

5. Add the times of First Quarter, Full, Last Quarter and New Moon, and perhaps the times of perigee and apogee. Your graph is then ready for use.

The diagram (Figure 4-1) shows the libration graph for the month of February 1982. Values of libration have been plotted every third day between the 1st and the 26th. The graph immediately provides the following useful information:

1. The NE and SW limbs were very well presented just before First Quarter and just before Last Quarter respectively. Unfortunately at both these times the best presented parts were in darkness and invisible!

2. At no time during the lunation were the NW and SE limbs well presented.

3. At the time of Full Moon (on the 8th) the southwest limb was reasonably well presented (the amount of tilt can be seen to be about 5°). Therefore, the thing to do was to look at the southwest limb on the evening of the eighth (just after Full Moon). There was little point in looking at the southeast limb on the night before Full since it was not well presented.

4. Plato was worth looking at as soon as it came into sunlight since the north limb was reasonably well exposed at First Quarter.

5. In the same way features in the southern hemisphere were well exposed at Last Quarter.

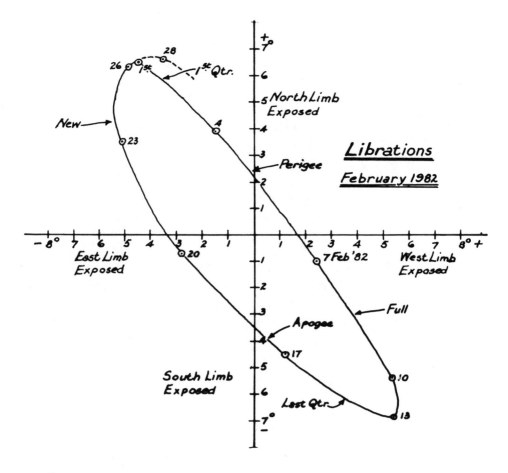

Figure 4-1. Librations. 1982 February.

The author has been plotting these libration graphs on and off for the past 15 years. So far as he can see, no two graphs are ever the same. They vary in shape, from almost perfect circles to very long, thin 'ellipses', rather like the February 1982 graph, only more so. It is these long ellipses which, when combined with suitable times of Full Moon, produce the choicest observations of the limbs.

So plot your libration graphs and make your plans accordingly.

THE SUN'S SELENOGRAPHIC COLONGITUDE

The Sun's selenographic colongitude is listed in the *Astronomical Almanac*, the B.A.A. *Handbook* and the Section *Circular*. The information is useful to the amateur astronomer, since it is equal numerically to the longitude of the morning terminator, measured eastwards (classical) from the mean centre of the disk. Thus, if the selenographic colongitude is 270°, the morning terminator will be at 90° W (classical), and it will be New Moon. The selenographic colongitude is approximately 270°, 0°, 90°, and 180° at New Moon, First Quarter, Full Moon, and Last Quarter, respectively. The longitude of the evening terminator differs from that of the morning terminator by 180°. At the morning terminator the Sun is rising, and at the evening terminator it is setting.

ABOUT THE AUTHOR

A navigator and hydrographic surveyor by profession, Henry Hatfield built his own 300 mm reflector to a unique design. With this instrument he has taken many photographs of the Moon, some of which have been published in his popular *Amateur Astronomer's Photographic Lunar Atlas*. A member of the B.A.A. for many years, he has served as a Council member and Papers Secretary. He is currently Director of the Instruments and Observing Methods Section.

5

SOME NOTES ON THE EARTH/MOON SYSTEM
by R.C. Maddison

The motion of the Moon around the Earth is often represented as comparatively simple, and yet there are numerous influences that make it, in reality, almost too complex to understand.

Most amateur observers are not much concerned over these knotty problems of theoretical dynamics, but they are fully aware of the strange irregularities and perturbations that lead to such phenomena as the librations, the tides, the captured rotation of the Moon, and so on. I have selected some of these more obvious phenomena to show how they can be understood in a simple qualitative fashion without getting involved in the detailed mathematics. What follows are summaries of some of the general properties of the Earth-Moon system and some simple explanations of the more important interactions of these bodies.

THE MOON

The Moon has a diameter of 3,476 km and has an almost perfectly circular profile. Careful measurements have, however, shown minor deviations. For example, the lunar surface facing the Earth is very slightly convex; it bulges towards the Earth by about 1 km. On the other hand, the whole lunar surface (bulge included) is depressed on the Earth-facing side and elevated on the side away from the Earth— the side we never see directly. In other words, if you draw a circle with the lunar north pole at the centre, then the whole lunar surface that faces the Earth would lie some 2-4 km below this circle—below the 'mean sphere' of the Moon—and the side facing away from the Earth would be between 1 and 5 km above the circle or 'mean sphere'; measurements from Apollo 15 and 16 have detected that.

The orbit of the Moon around the Earth is approximately elliptical, and varies in distance between 363,263 km (225,732 miles) at perigee and 405,547 km (252,007 miles) at apogee. This varying perspective is apparent on Earth as a variation in the angular size of the Moon of between $32'.9$ and $29'.5$, and this has important effects for solar eclipses.

The equatorial plane of the Moon is inclined to its orbital plane by $6°41'$, and this orbital plane is inclined to the ecliptic by $5°9'$ (Figure 5-1). These inclinations lead to a precessional effect similar to the Earth's, but, in this case, the rotation is much more rapid, having a period of only 18.6 years.

The mass of the Moon is measured to be 7.35×10^{22} kg (3.33×10^{22} lbs), which is, as a fraction $1/81.3$ of the Earth's mass (estimated at 5.977×10^{24} kg [2.71×10^{24} lbs]). This implies an average density of 3.34 g/cc (0.12 lbs per cubic inch), which is surprisingly low and indicates, when compared with the surface density, that much less of the mass is concentrated into a core than is true of the Earth.

ORBITS

The orbital motions of bodies in a system are related to the gravitational centre of the system. In the Earth-Moon system the barycentre (the point corresponding to the centre of gravity of the Earth-Moon system) lies within the Earth at a point some 1707 km (1060 miles) beneath the surface. This means that as the Moon rotates around the barycentre once a month, so also does the Earth, in an orbit of radius 4,671 km (2903 miles). In addition, both bodies are strongly affected by the Sun. For example, the Moon experiences extra gravitational force accelerating its motion while it is moving towards the Sun. On the other side of its orbit, while it is moving away from the Sun, the extra force slows it down.

The Moon follows the general laws of planetary motion around a centre, and so it moves faster in its orbit at perigee, when it is closest to the Earth, and slower at apogee, when it is furthest away.

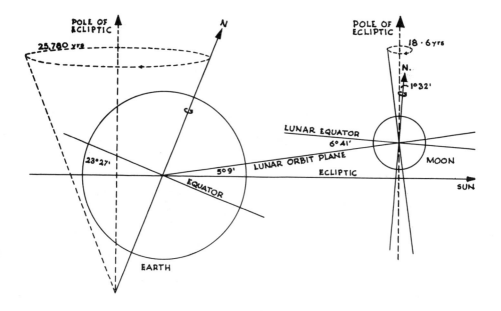

Figure 5-1.

THE LIBRATIONS

At any instant of time an observer on Earth can see rather less than 50 per cent of the surface of the Moon. This is simply because the limb that he sees from a single point is defined by tangential rays of light radiating towards his eye. The closer he is to the Moon, the smaller is the area of its surface that he can see.

If he watches the Moon for several weeks he notices that the limb is not always in the same place. Sometimes he sees features in one area that were beyond the limb a few days earlier, while at the same time features that were clearly visible on the opposite limb are lost to sight, over the horizon. The Moon appears to sway and nod as it moves along its monthly track around the sky. Careful mapping over a number of years shows that he can accumulate observations of up to 59 percent of the total lunar surface. The remaining 41 percent is, of course, hidden from view on the far side of the Moon.

The cause is lunar libration, mentioned in the previous chapter, which is due to three main causes. First, there is the libration in latitude, which accounts for the nodding motion. The plane of the Moon's equator is inclined some 6°41′ to the plane of its orbit, so that sometimes the Moon's north pole is tilted towards the Earth by this amount. At such times an observer on Earth sees slightly beyond the north pole, while the south pole is out of sight. Approximately half an orbit later, due to the nearly constant direction of the rotational axis relative to the stars, it is the south pole that appears tilted earthwards and the north pole that is out of sight. (Figure 5-2).

The second effect is the libration in longitude. In this case, the apparent swaying of the Moon is due to a mismatch of the constant rate of axial rotation and the varying orbital speed. Figure 5-3 is a plan view of the Moon's orbit. It shows the Earth at one focus and successive positions of the Moon at quarter-month intervals. It should be remembered that the Moon's axial rotation is quite constant, like a giant flywheel, and that it rotates through a quarter of a revolution in a quarter of a month. The orbital speed, on the other hand, is described by Kepler's Second Law of planetary motion, so that the Moon covers more of its orbit near perigee than it does near apogee in the same time interval. The resulting apparent sway to east and west of 6°17′ allows an observer to see alternately beyond the mean eastern limb and then beyond the mean western limb.

The third cause is a purely local one involving the rotation of the Earth. In half a day an observer is carried around half of the Earth's circumference. He may be able to observe the Moon at sunset over in the east and then to follow it across the sky until it is in the west at dawn. During this time his view of the Moon has changed because he sees it from different positions along a baseline of up to 12,000 km (7457 miles). This effect, small compared with the others, is called the diurnal effect.

CAPTURED ROTATION

When the Earth-Moon system originated some four-and-a-half thousand million years ago, it is fairly certain that the process involved was the gravitational capture of the Moon by the Earth and that the bodies were never linked together as some of the early theories supposed. If this was the case, then it is most unlikely that the period of axial rotation of the Moon was equal to its orbital

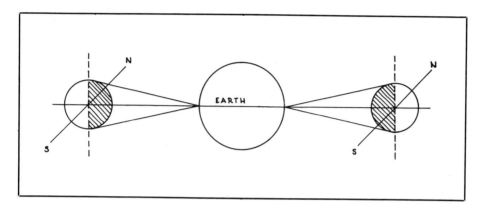

Figure 5-2.

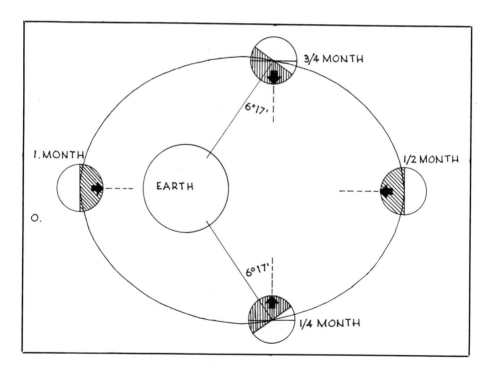

Figure 5-3.

period. The present situation is one that must have developed over a long period of time due to tidal interactions, and the chances are that the Moon's axial period was once much shorter than it is now.

The present captured rotation, whereby we always see the same face from Earth, is easy to understand if we remember that there is a tidal bulge near the middle of the Moon's visible face. The tide-raising force on the Moon is primarily due to the Earth, and one would expect the bulge to be formed in the direction of this force. In the early days, however, the Moon would have been rotating in a non-synchronous fashion, which means that the bulge would have had to move over the lunar surface in order to stay in the same direction relative to the Earth. There would have been a continuous plastic deformation of the Moon as it rotated underneath the bulge. This process would obviously involve considerable loss of energy because of the viscous drag of the bending rocks, and it would become progressively harder to move the bulge as the Moon cooled down and became rigid.

Over a long period of time this effect, which is like applying a steady braking force to a heavy flywheel, would slow the rotation down until the minimum amount of energy is lost in the process, and this would correspond to synchronous rotation where the bulge does not need to wander about over the surface. The same situation would have prevailed if the original period of axial rotation had been much longer than a month, but in this case the frictional drag would have slowly increased the rate of axial rotation.

In either case there would be a compensating effect on the Earth so that angular momentum is conserved, but there would be an overall loss of energy in the slight heating due to viscous drag.

THE RECESSION OF THE MOON

Another interesting side effect of the tidal deformation of the Earth is that it produces a gradual increase in the Moon's orbital speed which, in turn, makes the Moon slowly recede.

Figure 5-4 is a plan view of the Moon's orbit. It represents an instantaneous view of what is, in reality, a state of dynamic equilibrium. The Earth-Moon gravitation interaction generates the two bulges in the viscous but plastic body of the Earth, but the Earth has an axial rotation which is very much faster than the orbital rotation of the Moon. The effect of the frictional drag is that the bulges

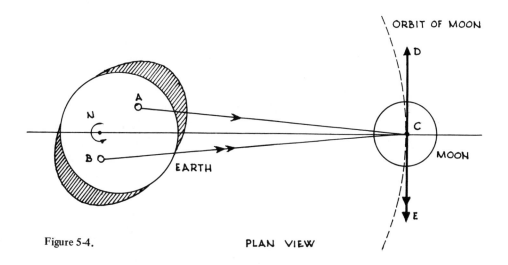

Figure 5-4.   PLAN VIEW

are carried around by the Earth's rotation until a balance is set up between the drag and the tide-generating force.

On Earth, an observer sees that the tides occur after the Moon crosses his meridian, almost as if the tidal bulges are anticipating the presence of the Moon.

One way of representing this situation in terms of forces is to regard the Earth as a double object having two gravitational centres, for example, at the foci of the ellipse produced in a cross-sectional view. As shown in Figure 5-4, at equilibrium, point A is nearer to the Moon than is point B and is therefore experiencing a stronger gravitational attraction than B. Both A and B are displaced from the central line, so that the forces along AC and BC converge towards the centre of the Moon. These two forces may be resolved at the Moon into components that act in one case along the central line towards the Earth, and in the other case at right angles into the direction of the Moon's orbit. The components acting towards the centre add together, whereas the components in the direction of the orbit are in opposition. Because the force along AC is larger than that along BC, there remains a net unbalanced force acting on the Moon in the orbital direction which has the effect of accelerating its motion. It thus moves into an orbit of larger radius, and modern estimates indicate a recessional speed of about 3.2 cm (1¼ inches) per year.

This effect is, of course, accompanied by a slowing of the Earth's axial rotation—since angular momentum has to be preserved—and there is indeed strong geological evidence that the days were shorter in the remote past than they are now, and there were more of them in a year.

ECLIPSES AND THE SAROS

One of the most helpful and fortuitous arrangements in astronomy is the near equality of the apparent angular sizes of the Moon and Sun as seen from Earth. On occasions the Moon's disk can exactly mask the brilliant solar photosphere, so that we can see the corona, the chromosphere, and prominences without specially designed equipment.

The varying distances of the Moon from Earth and the Earth from the Sun combine to give us a range of possible eclipses as the relative sizes of the Moon and Sun change. If an eclipse takes place when the Earth is at perihelion and the Moon at apogee, the Sun appears 1'35" larger in radius than the Moon, and the Sun is not completely obscured. Under these circumstances the Moon's shadow does not reach the Earth (Figure 5-5, case B), and a bright ring of the Sun is visible around the Moon. This 'annular' appearance can last for a maximum of 12 min. 24 sec. Conversely, when the Earth is at aphelion and the Moon at perigee (Figure 5-5, case A), an eclipse can occur where the Moon is 1'19" larger than the Sun and totality can last for as long as 7 min. 40 sec. The great African solar eclipse of 1973 (June 30) was close to this situation, the greatest duration at that time was 7 min. 5 sec.

In order to see a total eclipse one has to be at some point on the track of totality as the Moon's shadow moves across the Earth. Such tracks are at most about 250 km (150 miles) wide, and it is therefore not surprising that many of the Earth's inhabitants have never seen one.

On the other hand, a total lunar eclipse, where the Moon passes into the much bigger shadow of the Earth (Figure 5-5, case C), is visible to anyone in the appropriate hemisphere of the Earth. The size of the Earth's shadow is such that a total eclipse of the Moon can last for up to 104 minutes. Because of this, there is a distinct impression that lunar eclipses must be more common than solar eclipses. In fact, the opposite is true. Figure 5-5 shows how the area of sky on the sunward side of the Earth, in which a solar eclipse might take place, is considerably larger than the area covered by the Earth's shadow at night. This means that over a long period of time there are more solar eclipses than lunar eclipses.

Because the plane of the Moon's orbit and the plane of the ecliptic are not the same, eclipses do not occur every month. Figure 5-6

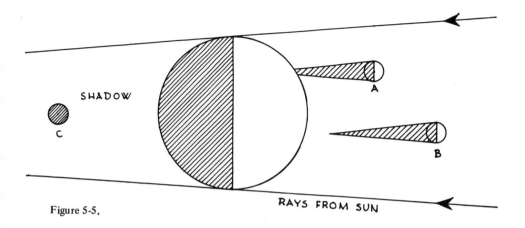

Figure 5-5.

shows the orientation of these two planes just before a solar eclipse. One of the nodes or points of intersection of the two planes must be in the direction of the Sun, so that as the Moon approaches the node it must also pass in front of the Sun. If these conditions can be satisfied, it can be shown that there may be between two and five solar eclipses in a year.

The prediction of eclipses is an art that was known to the ancient Babylonians, and it depended on a careful knowledge of the relative motions of the Sun and Moon and involved knowledge of the cycle of eclipses known as the saros. Solar eclipses, by definition, occur at New Moon, so the time interval between successive New Moons, the synodic month, is clearly an important factor. It was noticed by the early observers that during an interval of 223 synodic months the Sun completes a circuit of the sky to reach the same node at the same place on the ecliptic. This length of time is 6585.321 solar days, which is 18 years and $11^{1}/_{3}$ days.

The shortest time required for the Sun to travel from and return to the same node is 346.6 solar days, and this interval is known as an eclipse year. It is less than a calendar year because of the effect of orbital precession, mentioned earlier, which causes a slow regression of the nodes around the ecliptic. Nineteen of these eclipse years contain 6585.78 days, which is almost precisely 223 synodic months. The coincidence of these two periods means that, if an eclipse occurs at a node at a certain point on the ecliptic, it will be repeated at almost the same point after 18 years $11^{1}/_{3}$ days. This period of time is the Saros. In effect a cycle of eclipses repeats with this periodicity and, although the tracks of totality on the Earth's surface will not coincide, the appearances of the eclipses will be almost the same.

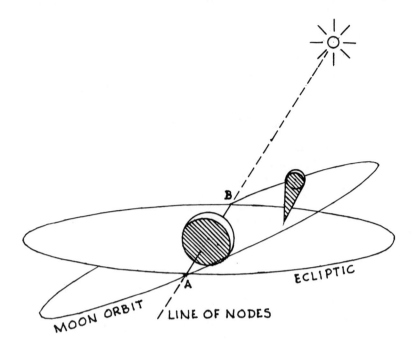

Figure 5-6.

If one takes account of the slight mismatch of 223 synodic months and nineteen eclipse years and calculates the exact period of the complete cycle, then the result is a series of eclipses lasting for about 1,200 years.

This brief summary of the more obvious properties of the Earth-Moon system show how complex the interactions are. When we bear in mind that the Moon is our closest neighbour, and that yet we do not fully understand its behaviour, we may realize how little we know of the rest of the universe. This is probably why astronomy is such a fascinating pursuit!

ABOUT THE AUTHOR

A former Director of the B.A.A. Lunar Section, past Vice-President and Council member of the Association, Ron Maddison is currently Senior Lecturer in Physics and Director of the Observatory at the University of Keele. He has a long standing interest in lunar studies, and his main interest is the promotion of astronomy through education. Other interests include the preservation of steam locomotion appliances.

6

DRAWING LUNAR FEATURES
by K.W. Abineri, with illustrations and
notes by L.F. Ball and A.V. Good

Using any given size of telescope, the human eye can observe far more detail than can be recorded on photographic film. Alternatively, this disparity can be expressed by saying that an observer using a 200 mm (8-inch) reflector can see detail that requires a 300 mm (12-inch) to record photographically. On the other hand, a photograph can portray a very much larger area and do so in a very much shorter time than it takes to make a drawing. However, the observer who can find enough time and patience to record what he can see in the form of a drawing, will get the utmost value from his telescope.

The new observer should attempt to draw features which are of special interest, starting of course with simple objects like single well-formed craters. This is the best method for really acquiring knowledge about the Moon's surface. It is excellent training for observational accuracy, especially at the present time since in many cases you can compare your completed drawing with photographs or charts and, in this way, assess your own ability to represent well-known lunar formations. Later, when you participate in the Lunar Section Programmes, your acquired drawing ability can be put to good use.

For those who wish to make drawings, the following hints may be of some value:

1. Use the minimum power necessary to show the visible detail. If the seeing is good, and there is much detail to record, do not try to draw too large an area.

2. During the observation, ensure that you are as comfortable as possible and, preferably, are seated. Have a suitable support for your drawing and a red lamp for illumination. These should be arranged in such a manner that drawing and observing may proceed without the necessity for moving away from the eyepiece.

3. Sketch in the outline of the formation and the larger objects. A detailed lunar drawing will take some time for completion, and during this period the shadows will change noticeably (an exception to this is in the polar areas, where the rate of movement of the shadow terminator is much reduced). It is therefore important to note the time when the outline and the shadows are drawn. Then proceed to draw the smaller details.

4. Outlines traced from photographs can be useful, but these may have to be modified to allow for differences in illumination or libration.

5. When making a sunset drawing of a large crater, record the detail near the shadow edges first, otherwise you will miss it. When making a sunrise drawing, record the detail remote from the shadow edges first.

6. Estimate the correct positions and relative sizes of all the objects in your drawing by careful comparisons at the telescope. If you are fortunate and possess a micrometer attachment at the eyepiece, make good use of this; however, with practice, eye estimates can be made quite accurately. Do not make the scale of the drawing too small.

7. Estimate the intensity of intermediate shades between dark shadows and brightly illuminated areas. If necessary, make notes about these. An arbitrary scale of '0' for dark shadows and '10' for very bright spots can be used and filled in on a separate outline sketch.

8. Make general notes during the observation and record the time when the drawing was started and the time when completed. Observing conditions must be noted, including details about the prevailing 'seeing' (definition, steadiness, transparency). Record the date of the observation, colongitude, etc.

9. Your drawing may be made with a soft pencil at the telescope (there are other methods), using light shading for the intermediate tones and heavy shading for dark shadows. Make all alterations and corrections at the telescope, using a good rubber eraser. Finally, compare your finished drawing with the telescopic view.

Afterwards the shadows may be painted over with Indian ink ("India ink" in the United States), and a light water-colour wash (or diluted Indian ink) can be painted over the intermediate shades. A mapping pen may be used to ink over fine detail. This drawing should be retained as the 'master copy' and marked with all the relevant information, i.e., period of observation, etc.

Any extra copies needed should be carefully traced from this original or photocopied.

10. Attempt to make your representation as objective as possible. Avoid all comparisons with other illustrations until after your work is completed.

11. When comparing your observations with the maps, photographs, and other drawings, remember the profound effects of varying illumination, libration, and seeing conditions, and avoid hasty conclusions. It is important to realize the limitations of visual work and to appreciate that all drawings, however accurate, contain subjective aspects. One observer may notice a small detail that is completely overlooked by another for no apparent reason. Furthermore, interpretations differ, more especially with regard to the finer detail.

12. It does take some courage to attempt a detailed drawing when the seeing conditions are really good, mainly because there is so much to be recorded, and one has a feeling of inadequacy; however, the new observer should not be deterred. The more drawings that you make, the more practiced you become, and each new observation should enable you to improve your technique. Also, visual acuity improves with practice.

ILLUSTRATIONS AND NOTES BY L.F. BALL AND A.V. GOOD

Figure 6-1 was made by Mr L.F. Ball, who has been drawing features of the Moon for many years. A description of his method follows:

> In order to obtain a reasonable outline, and if required, the immediate environs of the area one wishes to portray, I find it useful to check on available photographs including the photographic results of the latest lunar probes, particularly the Ranger and Orbiter series. Such an outline would, of course, have to be modified to conform with the existing libratory conditions by reference to the telescopic view.
>
> Whilst there are several ways of drawing the Moon, the most effective, though not the easiest, is the use of graphite pencils in various grades with

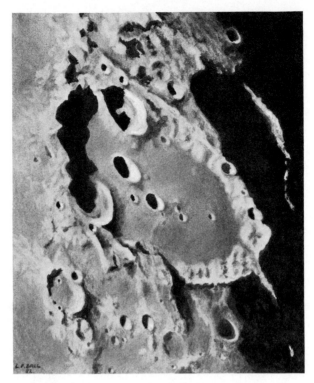

Figure 6-1. The lunar walled plain Clavius. 1982 March lunation. Moon's age: 9 days. 260 mm reflector, x250. (Drawing by L.F. Ball.)

Indian Ink for the shadows. Chinese White is sometimes useful for the highlights. This method, a favourite of mine, has been used for the drawing of Clavius. Line drawing is generally easier and quicker, but is lacking in aesthetic appeal.

I make quick sketches and notes at the telescope on a small drawing pad with a low powered flashlight bulb attached to the top of the board. These sketches are redrawn in permanent form as soon as possible after the end of the observation. Using the rough outline as a base, insert the most prominent shadows with a 4B or 6B pencil; the intermediate tones should then be put in with an HB pencil, and possibly a 2B for the heavier shades. Highlights should be carefully outlined and their relation to the dark shadows and intermediate tones carefully noted. Marginal notes are also very helpful. Most of mine refer to tonal intensities and the visibility and identification of minor details.

It is unwise to spend more than half an hour on any particular sketch, for obvious reasons. At first, this may be too short a time to do justice to the amount of detail revealed in a moderately sized telescope, but experience, as in all things, gives more confidence and speed. On the other hand, a more practiced eye sees much more and is able to take advantage of the moments of best 'seeing' conditions. One, therefore, has to compromise.

Preparation of the finished drawing should be made on smooth bristol board, the whiter the better. If the shapes of the dark and median shadows together with the highlights have been correctly drawn, the much sought after 3-D effect should begin to show, making the picture 'come to life', as it were.

Drawing the Moon is, without doubt, a fairly difficult exercise, taking into account the conditions under which the observer has to work, and one should not be too discouraged by one's early efforts.

Figure 6-2 was made by Mr A.V. Good, working directly at the telescope using a 2B pencil for the outline and a charcoal pencil for the shadow, deep shadow, and unilluminated parts beyond the terminator. The shading and deep-black areas were finished indoors directly after enough detail was recorded at the telescope. The finished work was made stable with fixative.

This drawing makes an interesting comparison with Figure 7-4, a photograph which includes a nearby area under almost directly opposite illumination.

Figure 6-2. Fracastorius. Conic Floor and Ghost Crater. 1968 February 3, 21.22 hrs U.T. 152 mm reflector, x180. Seeing: Ant. II. (Drawing by A.V. Good.)

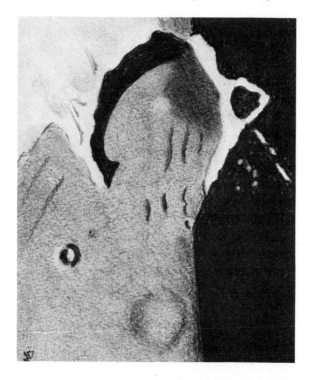

ABOUT THE AUTHORS

An industrial chemist by profession, Keith Abineri has been a member of the B.A.A. for many years. He has made many hundreds of lunar observations using a 200 mm reflector and was especially active during the period 1947-1960. His special interest is in the lunar limb regions and other selected features. Other interests include life and earth sciences, music, gardening and mountain climbing.

Using instruments ranging from a spectacle lens refractor to his present home-built 260 mm reflector and a recently acquired 203 mm Schmidt-Cassegrain, Leslie Ball's boyhood interest in astronomy led to a special interest in the Moon in his later years. A natural aptitude for drawing and painting has enabled him to depict lunar surface features in a lucid and artistic style. He is now a recognised expert in lunar cartography. A civil servant by profession, he has been a member of the B.A.A. for over thirty years.

By profession an architect, Arthur Good has combined his skill in draughtsmanship with a deep interest in the Moon to produce many fine lunar drawings. He has been a member of the B.A.A. for many years and currently serves on the Lunar Section advisory committee.

7

LUNAR PHOTOGRAPHY
by J.F. Pedler, with notes and photographs
by M. Mobberley, and photograph by J. Dragesco

Experienced lunar photographers have constructed unusual telescopes, cameras, shutters, and plateholders specifically for the purpose of taking photographs. Many amateur astronomers, however,
do not wish to go to such lengths, preferring to limit their photography to occasional sessions using a telescope that has been obtained
for ordinary visual work. These notes are intended for the beginner
who has a little basic standard equipment, but who wishes to take
good photographs. This is not to say that simple equipment may be
used in a rough and ready manner. On the contrary, it needs to be
used with much care and attention, with which it can produce
results—at times—comparable with those obtained using much more
sophisticated apparatus.

The very simplest method of taking a lunar photograph is to hold
a camera, focused at infinity, close to the eyepiece of a carefully
focused telescope. With a shutter speed of something like $1/25$ th
of a second, and a bit of luck, a passable 'snapshot' of the Moon may
be obtained. While the 'bit of luck' will always play its part in photography, it is to be hoped that these notes will provide the degree
of expertise that will make the bit of luck less necessary.

BASIC EQUIPMENT

The Telescope
It is very advantageous to have a reflector for use in photography.

A refractor is less convenient to work with and has two drawbacks for photography purposes. Firstly, it is rarely fully achromatic, which means that there will be a tendency for a slight 'fuzziness', particularly around the bright parts of the picture. Second, the visual and photographic focal planes of the refractor telescope will be in slightly different positions, with no guarantee that, having focused the image visually, it will be in focus on the emulsion. Filters can be used to correct this, but the exposure times will have to be increased, which brings further problems. It is preferable for the telescope to be equatorially mounted and to have a reliable drive in right ascension, although if exposures are kept below $1/30$ second, these are not essential. The type of focusing mount also is of some importance. As it is necessary to attach the camera to the focusing mount, it is highly desirable, if not essential, to have the rack-and-pinion type of focusing mount rather than the helical-screw type. (This latter method requires the whole camera to move in a cirular motion when being focused, and as the poor photographer will be trying to look through the camera at the same time, his head would need to assume very awkward positions.)

The Camera

In theory, all that the camera has to do is to hold the photographic film or plate in an exact position, ready to receive the light. Cameras can be simple boxes with the shutter just a hat thrown over the end of the telescope, but as this section is intended for the beginner with straightforward apparatus, this method will be left until later. The camera should be a standard 35 mm single lens reflex (SLR), provided that the lens system can be removed. With this type of camera, removal of the lens leaves a box containing the film and a shutter with variable speeds. It is even possible to buy a camera body without the lens to save costs. The great advantage of this type of camera is that what you see through the focusing aperture will, when the shutter is operated, be in focus on the emulsion.

The Adaptor

The third piece of apparatus is a means of linking the camera to the telescope focusing mount, so that the whole optical system is held rigidly in perfect alignment. Whatever method is chosen it should ideally allow the use of an eyepiece, a Barlow lens, or both, or neither. The cheapest, but not necessarily the simplest or best is a system of rods and clamps that will hold the camera body close to the eyepiece or empty focusing mount, but it will need to be well braced in order to prevent vibration. There is also the additional hazard of stray light (and there is plenty of it, especially with a near Full Moon) creeping in and fogging the photograph. Careful screening will be required to prevent this. By far the best method of mounting the camera to the focusing mount is by the use of a specially made adaptor. There are very many styles commercially available, with threads cut to fit most cameras, together with such eyepiece sizes as 25.4 mm push fit, 1¼″ standard R.A.S. thread, and 1¼″ push fit.

To be fully versatile, an adaptor ought to do the following:
1. Screw into the camera body.
2. Screw or slide into the focusing mount.
3. Allow your eyepiece to be inserted into it.
4. Allow a Barlow to be fitted.

In instances where either the camera or the eyepiece are of unusual size or thread, some suppliers offer the service of altering the adaptor to individual needs.

THE METHOD

There are two ways of photographing the Moon through the telescope, prime-focus photography and the eyepiece-projection method.

Prime-focus Photography

Full-disk photographs of the Moon can be obtained by placing the film at the prime focus of the telescope. Neither an eyepiece nor camera lens is needed for this type of photograph, but it will have to be determined that, by movement of the draw tube, the prime focus of the mirror can be brought outside the telescope. If this is not possible, then recourse must be had to a Barlow lens. If it is placed inside the normal focus of the telescope, the Barlow will intercept the converging cone of light and bring the focus futher out, where it becomes accessible to the camera and film. This method, however, gives an enlarged image (and a fainter one), requiring longer exposures. For a start, as experiments, the following range of exposures may be of some help, assuming a Newtonian reflector of approximately f/5 to f/7:

	Film speed	Shutter speed (secs)		
Full Moon	25 ASA	$1/8$	to	$1/15$
	64 ASA	$1/15$	to	$1/30$
	125 ASA	$1/60$	to	$1/125$
First or Last Quarter	25 ASA	$1/4$	to	$1/8$
(terminator region)	64 ASA	$1/8$	to	$1/15$
	125 ASA	$1/15$	to	$1/30$

Eyepiece Projection

Enlarged photographs of portions of the Moon can be obtained by using the eyepiece-projection method. As with prime-focus photography, the camera lens is not needed, the eyepiece of the telescope being used to enlarge the prime-focal image and project it onto the film. This is much like a slide projector that projects the image of a slide onto the screen, the degree of enlargement of the slide image depending upon the focal length of the projector lens and its distance from the screen. Increasing the distance enlarges the image on the screen but causes the image to become fainter, and vice versa. Similarly, if the distance from the eyepiece to the film is increased, the prime-focal image is enlarged but becomes fainter, giving rise to longer exposures and difficulty in focusing the fainter image. The focal length of the eyepiece will also have some bearing on the degree of enlargement. Use of a standard adaptor, as mentioned earlier, will give only a single degree of enlargement with each eyepiece used, since the distance between film and eyepiece is not adjustable. One way to alter this film-eyepiece distance is to change the eyepiece, though extension rings, which screw into the adaptor and are positioned between it and the camera, can also be used. Several will be needed, of various lengths. However, the construction of a bracket to hold the camera close to the eyepiece is possible, although great care is necessary to ensure that optical centres are exactly aligned and square on.

To determine the exposure time needed in this method of photography is not so simple. A series of photographs taken with different eyepieces at differing shutter speeds and with the eyepieces at differing distances from the photographic film will do more to give one knowledge of this type of photography than will pages of ex-

explanations or formulae. However, it will be a help to introduce two simple formulae.

If the focal length of the mirror (*FL*) is divided by the diameter of the mirror, then one obtains a figure called the (f)/ ratio (f), thus:

$$f = FL/D$$

For instance, a 150 mm (6-inch) mirror with a focal length of 1.2 metres (48 inches) gives

$$f = 1200/150 = f/8 \text{ (or, using inches, } f = 48/6 = f/8).$$

When using the eyepiece-projection method of photography it is necessary to calculate the effective focal ratio (EFR) of the whole system using the following formula:

$$EFR = f/ \text{ number of telescope} \times \left(\frac{\text{eyepiece to film distance}}{\text{focal length of eyepiece}}\right) - 1$$

It can be seen that with the focal ratio fixed, magnification can be increased by either increasing the eyepiece-to-film distance or by reducing the focal length of the eyepiece (using a higher power). Naturally, the bigger the EFR the fainter the image and therefore the longer the exposure needed. Typical exposures for a 212 mm (8½-inch) reflector (f/6.5) with an EFR of f/35 would be:

	Film Speed	Exposure time (secs)
First or Last Quarter Moon	25 ASA	1 - 2
(terminator region)	64 ASA	½ - 1
	125 ASA	up to ½

These figures are not to be taken as exact but merely as an indication. From them it can be seen that, almost always, an accurate drive is a necessity.

Taking the Photograph

After you have assembled the telescope and camera and decided upon the shutter speed, the actual mechanics of taking the photograph need to be considered. Whatever system is being undertaken, the biggest danger is that of vibration at the instant of exposure. The image of the Moon at the prime focus is quite bright, so exposures of a fraction of a second are used. To some degree this helps, but as it is the shutter and mirror mechanisms that are the cause of the trouble, a sturdy mount and a camera fixing device are the best help in overcoming the problem. It has been said before that the telescope drive need be used only for exposures of, say, $1/30$ second or longer. However, if a drive is available, it is convenient to use it to keep the image in the centre of the camera field of view while focusing or while taking a series of exposures. With the eyepiece-projection method the drive is essential, as exposures up to several seconds may have to be made. The vibration problem can be solved by the 'old hat' method. First, set the shutter to 'B' with the film wound on and the portion of the Moon to be photographed focused as sharply as possible. A lightproof lid, consisting of a cardboard disk, a thick black cloth or something similar, is placed over the end of the telescope. Ensure that the junction between the eyepiece and the camera is lightproof and, if the telescope tube is of the open variety, that no stray light can reach the emulsion. The shutter should now be opened and locked by the cable release. After allowing a few seconds for any vibration to die down, gingerly remove the lid from the end of the telescope tube. Estimate the length of the exposure either by counting or by listening to clock beats—and gently replace the lid. Unlock the cable release, which closes the shutter and completes the exposure.

THE FILM

The term ASA has been mentioned earlier and needs a word of explanation. It is an indication of the film speed or, more simply, how sensitive the film is to light. At this time there is no need to be concerned with how the values are obtained, but it must be understood that the higher the ASA number the faster the film, and therefore the shorter the exposure needed for any given light level. The term DIN sometimes appears and also ISO; these have a similar meaning. It may appear at this moment that the answer to most of the problems that beset the lunar photographer is to use faster and faster film, in order to have as short an exposure as possible. It is argued that vibration and seeing conditions would then have a minimum effect. While it is true that fast film may be used with some advantage when no subsequent enlarging is intended, at other times it will be a real disadvantage. This is due to what is called grain. Emulsions consist of minute particles of a silver salt embedded in a thin film of gelatine. The grains of the silver salt in a fast film are comparatively large, with a tendency to clump together, and although they respond to faint light, they lack resolving power and will spoil fine lunar detail. A slower film needs much more light but its resolving power will be very much higher. For the beginner it is better to use a medium speed film, around 64 ASA, until experience has been gained.

The question now arises as to the choice of colour or black-and-white film. To a certain degree this is a matter of choice, but several points should be borne in mind. By and large, home processing of colour film is somewhat specialized. What is more, though all films have a certain 'latitude' whereby a range of exposures will yield a satisfactory image, the latitude of colour film is very small; there is very little room for error. It should also be remembered that, apart from lunar eclipses, there is very little colour on the Moon. In fact, different brands of colour film will impart their own suggestions of colour to the lunar surface. Also, colour film is more expensive than black-and-white photography, if slides are chosen then film can be obtained that gives a positive image direct, while if film for prints is chosen, a whole array of techniques of developing and printing is available.

SOME USEFUL THEORY

So far, almost without exception, formulae and theory have been deliberately omitted. At the risk of repetition, it is the writers' opinion that straightforward trial and error will, in the initial stages, teach the beginner far more than pages of often incomprehensible equations. However, no introduction to the subject of lunar photography would be complete without some theory, as given below.

Irrespective of size of mirror, all telescopes with the same f/ number will have the same image brightness at the prime focus of the telescope. This image brightness varies inversely as the square of the f/ number. An image at f/16 will be four times fainter than one at f/8 and will require four times the exposure. For any given diameter of mirror, a longer focal length will produce a larger f/ number and a larger and fainter image (and require a longer exposure). These remarks apply equally well to the eyepiece-projection method, and a formula that will be found very useful is:

If M is the desired magnification of prime focal image,

    FL_e is the Focal Length of an eyepiece

    D is the distance between film and eyepiece

then $(M + 1) FL_e = D$

For example

if $M = 5$ and $FL_e = 12$ mm (eyepiece)

then $(5 + 1)12 = 72$ mm (between eyepiece and film).

In this example the effective focal ratio (EFR) will be 5 × the focal ratio of the telescope, and the effective focal length (EFL) will be 5 × the focal length of the telescope.

If the image diameter is measured from a previously made trial exposure, or by measurement of a white card held up at the eyepiece at the correct eye-piece film distance, then the EFL in the units preferred (mm or inches) is given by:

$$\text{EFL} = 206265 \times \frac{\text{Image diameter in preferred units}}{\text{Angular diameter of subject in seconds of arc}}$$

If, for example, the measured image was 2.5 cm and the Moon's diameter is taken as 31 × 60 seconds of arc then:

$(20625 \times 2.5)/(31 \times 60) = 277.24$ cm which is approx. 2.8 metres.

Finally, the ultimate formula for exposure is:

$$t = \frac{KN^2\ w\ (\text{or } y)}{BS}$$

Where

    t =exposure time in seconds

    K =36 (constant)

    N =EFR of the telescope (will be the f/number with focal plane photography)

    w =luminance of the Moon according to phase—see Table

    y =luminance of the Moon according to phase—see Table

    B =lumens per square metre constant at 7760

    S =film speed in ASA

For example, at the focal plane of the telescope, what exposure is needed to photograph the terminator of the seven day Moon with an f/8 telescope and a film speed of 32 ASA? Known quantities are:

$K = 36$ $S = 32$ ASA

$N = 8$ $y = 6.9$ (from the Table)

$B = 7760$ (Lumens/square metre)

Evaluating: $(36 \times 8^2 \times 6.9)/(32 \times 7760) = {}^1/_{16}$ second (approx.)

With the Moon at an altitude of less than $30°$, exposures should be doubled at least to allow for absorption of the Moon's light by the Earth's atmosphere.

TABLE

Age of Moon (days)	w (used for limb regions)	y (used for terminator)
14	1.0	1.0
13 and 15	1.4	1.4
12 and 16	1.8	1.8
11 and 17	2.3	2.4
10 and 18	2.9	3.2
9 and 19	3.5	4.1
8 and 20	4.2	5.3
7 and 21	5.1	6.9
6 and 22	6.2	9.4
5 and 23	7.4	13.0
4 and 24	8.7	18.0
3 and 25	10.5	34.0

AFTER THE EXPOSURE

Normally, 35 mm film is bought in rolls of 20 or 36 exposures. Once every frame on the roll has been exposed, the film has to be developed so that the latent image can be seen. In the author's opinion, beginners should have their films developed commercially until some experience has been gained. If the film is for 35 mm slides, then development will be all that is required, since the returned slides will be ready for showing in a projector. However, if the film is black-and-white negative film, prints will need to be obtained. (If the negative film is 35 mm, the prints will have to be enlargements.) Probably the best way is to have the film developed and then to select only those images that will give a good enlargement. However, it has to be admitted that commercial work will become expensive if more than the occasional film is used.

Anyone who takes up the hobby of astronomical (not just lunar) photography will need to enter the realms of developing and printing. The subject needs more space than is available here, but a few suggestions will not be out of place. The most convenient way to develop 35 mm film is in a developing tank. Tanks are advantageous because once a film has been inserted (no mean feat in total darkness), all subsequent operations can be done in daylight. A short list of basic equipment should include the following:

Developing tank to suit 35 mm film, preferably 36-exposure size.
Developing chemicals to suit the film being used.
Stop bath solution (universal).
Fixing solution (universal).
Wetting agent (universal).
Thermometer.
Clean running water.
A room that is perfectly dark or a lightproof bag or box.

The darkroom is needed for loading the film into the developing tank. The lightproof bag is mentioned as a last resort, since it is virtually impossible to keep it free of the photographer's worst enemy, dust. (Incidentally, the very worst place to load film is under the bedclothes.) Unless the reader intends to pursue photography very actively or regards expense as no object, then it is not worth fitting up a room especially as a darkroom. For normal use, the bathroom is probably the best place, as it will be relatively dust free and contain a supply of hot and cold running water. It should not be difficult to make a frame of lightproof material to fit the window, and a thick curtain hung up at the door will be good enough to obstruct light from that source.

After the film has been loaded into the tank (practice with a length of old film first), the procedure is a fairly straightforward affair. All film, chemicals, and apparatus should be at a stable, reasonable temperature, say about 20°C (68°F). The length of developing and fixing times should be strictly controlled according to the manufacturer's instructions, and the method of agitating the solutions while in the tank should also be standardized. After the process has been completed and the film removed from the tank, it should not be wiped, but hung up to dry naturally. If enlargements are necessary, further apparatus is necessary: an enlarger, printing papers, and one or two small items for finishing the prints.

All good photographers have one habit in common: they keep extensive records of all photographs they have taken, whether good or bad. Notes of exposure times, seeing conditions and development details are all to be recorded, as these form an invaluable source of reference for the future. The motto should be: record everything.

CONCLUSION

It is hoped that these notes have shown how the owner of a small or moderate-size telescope can take good and pleasing photographs of the Moon. The degree of involvement depends only upon the interest of the individual. He may decide to use only commercially available products or he may opt for making all of his components himself. Both ways will, used properly, give good results. The illustrations show what can be achieved.

SOME SUGGESTIONS ON TAKING PHOTOGRAPHS
(from M. Mobberley)

Equipment

A most appropriate telescope is a large amateur Cassegrain such as the author's 350 mm (14-inch), with the facility for eyepiece projection from f/60 up to f/200. This was used to take Figures 7-1, 7-3, and 7-4. Figure 7-2 was taken with a slightly larger telescope.

A good camera is the Olympus OM-1, a camera with a very low shutter vibration, a reflex mirror (which should be locked in the up position after focusing on to the focusing screen), and an air shutter release for avoiding shutter vibration.

All equipment must be rigidly mounted and the polar axis very accurately aligned.

Figure 7-1.   Ten-day old Moon. 1981 December 6 at 21.08 hours U.T. 1/500-second exposure at f/5 on XP1-400. (Photograph by M. Mobberley using 350 mm (14-inch) Cassegrain.)

Figure 7-2.   Mare Imbrium. 1982 January 3 at 19.10 hours U.T. Celestron 14, F/D=40, on XP1-400. (Photograph by J. Dragesco).

Figure 7-3. Mare Crisium. 1981 November 13 at 22.51 hours U.T. 1/60-second exposure at f/20 on XP1-400. (Photograph by M. Mobberley.)

Figure 7-4. Region around Theophilus. 1981 October 16 at 23.15 hours U.T. 1/8-second exposure at f/34, 25 mm eyepiece projection at Cassegrain focus, on XP1-400. (Photograph by M. Mobberley.)

Focal Length of the Telescope

The longer the focal length and the higher the f/ ratio the better, providing that the exposure time does not become too long for the prevailing seeing conditions (turbulence).

Focusing is often very critical at low f/ ratios. It is 16 times easier to get the film plane exactly right at f/20 than it is at f/5, NOT just 4 times.

Preparation for Photography

Before starting photography, have the observatory open for as long as possible. This improves seeing, since it reduces turbulence by allowing internal and external temperatures to match up (and so, improve resolution).

Remember that body heat, breath, low wattage red bulbs, and air currents, both within a closed telescope tube and within the observatory (particularly the dome), all cause turbulence.

Taking the Photographs

Ten seconds before taking photographs, move as far away as possible from the telescope and stop breathing!

When seeing conditions are good, use film generously, up to 36 exposures on any chosen subject. Develop the film as soon as possible and examine the negatives with a magnifying glass. Print only the best photograph. (There will usually be one exposure made when the atmosphere was least turbulent.)

In general, the best times for photography are when the Moon is high in the sky. The higher the Moon the better the seeing, i.e., excellent atmospheric transparency with least turbulence and spurious colour.

Seeing conditions often deteriorate when temperatures drop below the freezing point, though atmospheric transparency usually improves. Experience has shown that the best nights are usually those when the atmosphere is humid and misty near the horizon.

Films to Use

Good films are Ilford XP1-400, Kodak Technical Pan 2415, and Ilford Pan F. These are usually used at 400, 100, and 25 ASA respectively. My personal preference is for XP1-400, due to its tolerance to overexposure.

References
American readers can find useful references in the Eastman Kodak Company's booklet AC-48, *Astrophotography Basics*, and in the *Photographic Exposure Guide* (ANSI PH 2.7) of the American National Standards Institute. There is also Dr Donald Parker's exposure system given in the 1980 June-July issue of *The Astrograph*.

ABOUT THE AUTHORS
A member of the B.A.A. since 1969, Martin Mobberley built his own 210 mm reflector in 1973 and now observes with a 350 mm reflector. Astronomical interests include lunar and planetary photography, comet sweeping and monitoring the lunar surface for transient phenomena. He is currently co-ordinator of the Lunar Section's photographic sub-section and a member of its advisory committee.

An amateur astronomer with many years experience, John Pedler observes with a 310 mm reflector from his home in Bristol. He has a special interest in the study of lunar surface detail. He is currently Librarian of the Lunar Section and a member of its advisory committee.

8

PHOTOGRAPHING LUNAR ECLIPSES
by M. Mobberley

Eclipses of the Moon, total or otherwise, are relatively rare events on the astronomical calendar, and so when one occurs, weather permitting, it is a good idea for the budding eclipse photographer to be well prepared. Not only should the observing equipment be overhauled prior to the event, but a check list and schedule should be drawn up detailing the proposed exposure times for each stage of the eclipse and the 'plan of action'. Even if the sky is cloudy at the start, the observer and his equipment should be poised for action in case the sky suddenly clears. If possible a spare set of equipment should be available at the telescope. All this forward planning might seem a bit excessive, but, in the author's experience, if something is going to go wrong it will invariably do so in the heat of a big event. Everything from ball-point pens to camera shutters can go wrong, and even the most patient observer will not relish the prospect of waiting several years for the next total eclipse, with no guarantee of a clear night even then. It is a good idea to have loads of film on hand, all legibly marked with the film speed to avoid confusion in the dark. A large amount of film allows the photographer to take several shots at a variety of exposures during each stage of the eclipse, to be sure of obtaining the best possible picture, and to guard against flaws in the emulsion. If the films are to be developed commercially (which is often the case with colour film), it is a good idea to send alternate cassettes to different places as a further safeguard. Those amateur astronomers who prolong the lifetime of their film by keeping it in a freezer should allow it to thaw out for at least five hours prior to exposure, to avoid condensation problems.

Precautions apart, what plan of action should the eclipse photographer adopt? A lot depends on the equipment used. The optimum instrument for lunar eclipse photography is a big, fast reflector with a focal length of less than 90 inches (2.3 metres). From the point of view of light grasp a Newtonian reflector is most efficient at about f/4 for a given size of secondary mirror, so the ultimate Newtonian for lunar eclipse photography would be one of about 20 inches (0.5 metre) aperture and f/4 focal ratio. The author's own 14-inch (0.35 metre), f/5 Newtonian proved to be very versatile during the 1982 January 9th total eclipse. Figures 1, 2, and 3 were taken with this telescope. Kodacolor 400 was used throughout. A focal length greater than about 90 inches (2.3 metres) will mean that the Moon will exceed the frame size of the popular 35 mm format film.

There is real scientific value in timing the passage of the umbral shadow across certain lunar features. This is usually done visually, but photographs of those events are also desirable. However, the latter are not easy and require a large and photographically 'fast' telescope.

The main problem facing lunar eclipse photographers is the huge range of brightness levels encountered, especially if colour film is used. The human eye is able to adapt well to large variations in brightness, but colour film is not. With very transparent skies and 400 ASA (27 DIN) film at f/5, the Full Moon requires an exposure of about 1/1000th of a second. This is the minimum exposure available on most SLR cameras. However, during totality the Full Moon can become almost invisible to the naked eye, dropping from magnitude -12.7 to magnitude +4 in extreme cases (such as the 1982 December 30th eclipse). This is a reduction in brightness of some five million times! Although eclipses as dark as this are rare, they do occur, and at f/5 even exposures of many minutes with sensitive film will only pick out a few details on the lunar surface. Really dark eclipses are usually caused by dust in the Earth's upper atmosphere and to a lesser extent by the centrality of the Moon within the umbra. Ultra-fast Schmidt cameras, with f/ ratios of about 1.5 and hypersensitized film will be needed to record the darkest lunar eclipses well. With exposures of many minutes, fast film is not necessarily an advantage as what is known as 'reciprocity failure' tends to make nonsense of normal film speeds. In fact after hypersensitizing film in a 'cold camera' or with gas treatment before use, slow film sometimes becomes more sensitive to light with long exposures, than fast film. Those amateurs with Schmidt cameras will be well equipped to deal with very dark lunar eclipses. However, those Schmidt cameras within the range of most amateur astronomers have focal lengths between 9 and 12 inches, (230 mm and 300 mm), which will produce a relatively small lunar image on the film. Therefore, relatively slow film should be used with these cameras to permit enlargement of the image without excessive grain. Unfortunately most amateurs do not possess Schmidt cameras, or use hyper-sensitisation, so the following advice has been prepared for amateurs with more conventional equipment.

At the start of any lunar eclipse the penumbra of the Earth's shadow begins to creep across the lunar disk at about half a mile per second. At this stage of the eclipse the required exposure, to portray the Moon as seen by the observer, will be the same as that required to capture the Full Moon, i.e. about 1/1000th of a second at f/5 on 400 ASA

Figure 8-1.   Twenty-three minutes before first umbral contact. Exposure 1/1000th of a second.

film. At such short exposures the normal relationship between exposure time and f/ number will apply: that is if you halve the film speed you will need to double the exposure time. If your telescope is working at f/10 you will need to use four times the exposure which you needed at f/5. So far this is all fairly simple, but when photographing astronomical objects with standard daylight film there is often a large range in the exposures which different films require. Amongst the 400 ASA films, Kodacolor 400 is the fastest, black and white Ilford XP1-400 lies in the middle, and Ektachrome 400 is generally the slowest, when used on the Moon. However, Ektachrome does give a very true rendering of the colour of the lunar surface whereas Kodacolor tends to produce a slightly green Moon. Because of this variation between films, and the added factor of atmospheric transparency on the night, it is a good idea to 'bracket' exposures, i.e., take other photographs at half and double the original exposure. With an f/5 telescope photographing the Full Moon in colour, you really need to use 200 ASA (24 DIN) film at 1/1000th, 1/500th, and 1/250th of a second exposures to safeguard against under- and over-exposing the image. Black and white film has a much greater latitude and so one exposure will often suffice. A yellow filter can be used to advantage with black and white film on the Full Moon to reduce glare.

With the eye, even using a telescope, the penumbra is quite difficult to detect for the first ten minutes, as at the edge of the penumbra only a fraction of the light is actually being cut off. However, by underexposing the Moon, the penumbra can be enhanced and successfully recorded on film. Ektachrome 400 is not too fast for this: 1/1000th of a second at f/5 can be used to detect the penumbra with this film. The eclipse photographer should not worry about confusing his umbra photographs with penumbral ones as the penumbra is unmistakable due to its larger radius and hence straighter edge.

Once the Moon is well into the penumbra, the photography starts to become more difficult. The eclipse photographer has to choose between photographing the unaffected part of the Moon at the normal Full Moon exposure, or photographing detail well within the penumbra. In practice the best compromise is to wait until the penumbra is more than halfway across the disk, and then double all your exposures right up to the point where the penumbra is covering the whole of the Moon. At this stage the umbra will start to bite into the Moon, which is when the problems really start. A well-prepared photographic schedule becomes invaluable from this moment on.

Figure 8-2. Five minutes after third umbral contact. Exposure 3 seconds. The edge of the umbra was relatively bright.

Figure 8-3. Twelve minutes before fourth umbral contact. Exposure 1/500th of a second.

As the umbra moves across the Moon's disk the observer will notice that it becomes harder to read his schedule, or to make notes, by moonlight, and so a dim red light will be useful at this point. The observer's schedule should give details of the proposed exposure times at each stage of the eclipse, and be designed to avoid changing films during critical times (e.g., just before and after totality when the most spectacular views are obtained).

What exposures should be used once the Moon is within the umbra? In general, only very short or very long exposures will give good results. Very long exposures will bring out detail in the umbra, but grossly overexpose the penumbral detail, whereas very short exposures will leave the umbral detail black, but show the penumbral detail reasonably well. Intermediate exposures will overexpose half the image and underexpose the rest, producing a picture with no lunar surface detail visible. The best policy is probably to treble the normal full Moon exposure until the umbra is at least three-quarters of the way across the lunar disk, and then switch to very long exposures to reveal detail within the umbra. If very long exposures are used before the umbra is at least three-quarters of the way across the disk, the glare from the bright limb will tend to swamp the whole picture.

Judging the length of exposure to capture detail within the umbra is the most difficult task facing the lunar eclipse photographer. Unless he/she is using a very slow system of f/10 or greater (in which case exposures of many minutes will be required and film speed will be irrelevant), very fast film is essential. At the time of writing (1983), both Kodak and 3M have just introduced 1000 ASA (31 DIN) colour film, and this is the sort of speed required to capture umbral detail in seconds, rather than minutes. Film of this speed should resolve about 80 lines to the millimetre, which translates into a resolution of better than 2 seconds of arc (or 2 miles on the Moon) when used with a reflector like the author's 14-inch (0.35 metre) Newtonian at f/5. This sort of resolution is not much worse than the limit imposed by the Earth's atmosphere on a normal night.

Amateur astronomers with slower film or slow optics will be forced into exposures of many minutes which, even with a good commercial drive system, will necessitate guiding. Because the Moon drifts relative to the background stars at about half a second of arc every second of time, the guiding must be done on the lunar surface, rather difficult during totality, or by calculated offset from nearby stars (which must not be east of the Moon to avoid occultation). The popular off-axis guiders now available do produce a rather dim, distorted image and, if possible, a separate guide telescope is preferable. It is far easier to use fast optics and fast film, as guiding can be very tedious, and will probably not produce a sharper image even than 1000 ASA film when used with a large amateur reflector (guiding to better than 2 seconds of arc is almost impossible for any amateur to achieve).

Providing the eclipse is not exceptionally dark (i.e., 0 or 1 on the Danjon scale), the exposures required with 1000 ASA film and fast optics, to record umbral detail, will be similar to those required for capturing Earthshine, i.e., up to about 10 seconds at f/5. Most good amateur drives will track within a few seconds of arc at the lunar rate for 10 seconds, and so umbral detail is easily recorded. However, it is virtually impossible to judge the required exposure during totality, so a variety of exposures, up to 10 seconds duration, should be attempted. It will be found that background stars down to about tenth magnitude can be recorded at f/5 with 1000 ASA film and a 10 second exposure. In 10 seconds the Moon will drift about 5 seconds of arc eastward with respect to the background stars, and it frequently drifts in declination too; at worst it may amount to 15 arc seconds per *minute* of time. These are good reasons for avoiding long exposures: they will produce noticeable star trails. After totality the required exposure times will be the same as before, provided the atmospheric transparency does not change.

For those observers without access to a telescope, a fixed camera with the facility to take multiple, or very long exposures, can be used to good advantage to record the passage of the Moon through the Earth's shadow. Slow film is best for this type of photography as the lunar image will be small, and most modern cameras have very fast lenses. Whether using a single lens reflex camera for lunar eclipse photography or on other astronomical subjects, it is a good idea to change the standard focusing screen for a clear one to assist in photographing dim objects. Some really spectacular shots of lunar eclipses over landscapes can be taken with a bit of careful planning. Exposure details for these types of shots are best gleaned from relevant articles in the popular astronomical magazines and periodicals. Perhaps the ultimate in lunar eclipse photography is to use a filtered, 'piggy-backed', telephoto lens of about 400 mm (16-inch) focal length, with a field of view of a few degrees, guided on the background stars throughout the umbral stages of the eclipse. While the observer guides through the telescope on a fixed star, the lens records the Moon's passage through the umbra, and hence photographs the profile of the Earth's shadow. If a portable telescope is transported to the right observing site, spectacular grazing occultations by the eclipsed Moon can be photographed, showing a star disappearing behind the mountains on the lunar limb. Such photographs are of scientific value as they can be used to determine the lunar diameter, the Moon's position in orbit, and the profile of the lunar limb, with great accuracy. However these sort of photographs require considerable forward planning, and are only usually attempted by very experienced amateurs.

To summarize then, any amateur with suitable equipment (preferably a fairly fast Newtonian), can, with care, take really good lunar eclipse photographs, providing he plans his exposures well, and does not mind being a little extravagant with his film. Although not as spectacular as a solar eclipse, a lunar eclipse is a rare and fascinating event, and well worth recording photographically.

ABOUT THE AUTHOR

A member of the B.A.A. since 1969, Martin Mobberley built his own 210 mm reflector in 1973 and now observes with a 350 mm reflector. Astronomical interests include lunar and planetary photography, comet sweeping and monitoring the lunar surface for transient phenomena. He is currently co-ordinator of the Lunar Section's photographic sub-section and a member of its advisory committee.

9

THE OBSERVATION OF
TRANSIENT LUNAR PHENOMENA
by P.W. Foley

INTRODUCTION

The term 'Transient Lunar Phenomenon', TLP, is used to define a visual observation of the lunar surface whereby the observer has encountered an abnormal appearance of a feature.

The Moon, our nearest neighbour within the Solar System, has for centuries puzzled many by an ever-changing spectacle. During the course of a lunation, any given point on the surface will undergo tremendous change while viewed from Earth. A crater will exhibit shadow and be seen in relief during periods in the lunar forenoon and afternoon. However, this same crater will become completely devoid of shadow for a while prior to and after local noon and may well merge almost indistinguishably with surrounding terrain during this period of high angle of illumination (Figures 9-1 and 9-2). Is it any wonder, then, that observers have reported abnormalities that can now be attributed to natural effects directly associated with angle of illumination? Yet, there still remain reports from the past and, indeed, from the present, that cannot be dismisssed in such a fashion. Notables from the past, Herschel, Schröter, Bode and Elger, to name just a few, recorded abnormalities. On one occasion Herschel described the crater Aristarchus in dramatic terms, "like a nebula—like glowing coals". It is to more recent times that we must turn for positive proof that events witnessed are not effects conjured up by Earth's atmosphere; photographic, photometric, polarimetric, spectrographic, and spectrometric records now exist. On 1969 April 1st, professional astronomer N.A. Kozyrev at the Crimean Astrophysical Observatory obtained a spectrogram of a red spot on the inner western wall of Aristarchus from which CN molecules and N_2 emission were identified. While orbiting the Moon Apollo astronauts were witness to events. Armstrong, Aldrin, and Collins verified a brightening in Aristarchus which was observed simultaneously by independent terrestrial observers. Schmitt (1972), on Apollo 17, saw a flash in Grimaldi, and Evans, also on Apollo 17, saw a flash in Mare Orientale. Evans did not see any cosmic ray flashes in the whole trip, whereas Schmitt did. He said he was dark-adapted at the time and could not say whether it was a cosmic ray flash or lunar event. Many historical observations, however, have been of flashes in and near Grimaldi. Therefore, it is highly likely that Schmitt observed a lunar event. Mattingly (1971), on Apollo 16, also observed a flash when he was on the far side of the Moon, but it was below his horizon and he could not identify the region. Therefore, permanent records and astronaut sightings with no atmospheric interference corroborate the visual reports of transient lunar phenomena.

NASA have done much to encourage research into the subject, and it is thanks to the efforts of Winifred Cameron (USA), Barbara Middlehurst (USA), Patrick Moore (UK), and others that we now have much documented material. It is from catalogues by Winifred Cameron and Patrick Moore that the foregoing historical notes have been extracted.

Within the Lunar Section of the British Astronomical Association exists a Sub-Section the aims of its members being to observe and confirm phenomena and to record and tabulate the findings. This Sub-Section works in close harmony with its American counterpart; more information is given at the end of this article. Although these networks are essentially amateur organizations, much help and advice is given by professional bodies. The following paragraphs are intended as a guide to those who may wish to participate. A high standard of observational skill is required from those taking part, but new members are always welcomed and given every encouragement to obtain the necessary expertise.

THE INSTRUMENT AND THE OBSERVER

What type of telescope is suitable? The answer here is simple. Any type will do, provided the aperture is adequate. In the case of reflecting instruments, a mirror of 212 mm (8½ inches) is probably the

Figure 9-1. The three large craters are, from the top of the picture to the bottom, Arzachel, Alphonsus, and Ptolemaeus. Alphonsus is distinctive by way of the dark markings upon its floor and is a TLP site. The photograph was taken when the angle of illumination permitted shadows to form relief. (South is to the top of the page.)
Courtesy of the Lunar and Planetary · Laboratory University of Arizona.

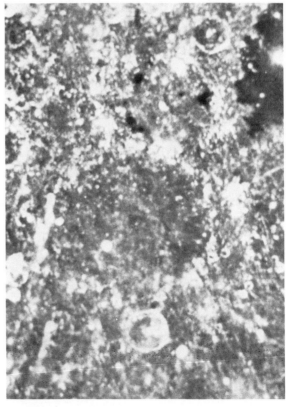

Figure 9-2. The area covered by this photo is nearly the same as by Figure 9-1 except that Arzachel is not shown. It was taken when the angle of illumination was high over the region and all relief eliminated. Alphosus is easily identified by the dark floor markings. (South is to the top of the page.)
Courtesy of the Lunar and Planetary Laboratory University of Arizona.

desirable minimum, but good results have been obtained from instruments of 150 mm (6-inch) diameter. A lot will depend upon the quality of the mirror. Currently in use with the Section are aperture diameters from 150 mm to 450 mm (6 to 18 inches), and quite obviously the larger mirrors do have greater resolving power and light grasp. Refracting instruments should be 112 mm to 125 mm (4½ to 5 inches) minimum; the refractor will not, however, have as good response to colour as a reflector.

Most important, of course, is the observer. He or she will have to have dedication and determination to be willing to spend many hours at the telescope. Just looking at the Moon is not good enough; detailed examination is vital, and a thorough knowledge of any region under surveillance, under all angles of illumination, is a must. A good atlas or lunar map is a great help, a photographic atlas invaluable. Another must: never go to the telescope without a pencil and some paper.

OBSERVATIONAL METHODS

Two methods of observation have been adopted. One is scanning, with the hope of detecting at random, some abnormality. The other is to do surveillance of specific regions.

General scanning has been recognized as rather a hit-or-miss affair. On the other hand, study of specific regions has proved to be highly successful.

It is unlikely that one individual can gain complete visual knowledge of 'our' side of the Moon. The best strategy—certainly for the beginner—is to be selective and observe a limited number of areas. The Section suggests a maximum of six to start with, including three known TLP sites and another three of the observer's personal choice.

RECORDING OBSERVATIONS

The very essence of TLP detection is accurate and precise reporting of any observation undertaken. To meet this requirement, the Section has adopted a standard report form. Figure 9-3 is an example of a routine report. It is equally as important to record routine observations as those that may be thought to show phenomena. Very often it is necessary to refer to previous work in order to clarify points that arise at a much later date; the motto is, never waste an observation by omitting to record it. Elaborate sketches and long, detailed descriptions are unnecessary. As our example shows, considerable information can be imparted easily, in quite a brief manner. All records should contain the name and address of the observer. Never forget the date; always use the astronomical format, that is the year first, followed by the month, and then the day of the month. Include details of the instrument used and magnification used. It also is helpful to include information on local conditions: temperature, a word or two about the weather, and if possible, the barometric pressure. Any observation that does not carry time may be wasted; use Universal Time, abbreviated as UT. Make reference to any accessories employed. Always be brief and accurate. When phenomena are encountered always ensure that comparative studies are made of nearby and similar features. This action is vital in order to establish that the effect seen and considered abnormal is not widespread.

SEEING CONDITIONS, TURBULENCE, AND TRANSPARENCY

Turbulence within the atmosphere of Earth will be seen as a rippling (some call it 'boiling'), and when viewing the lunar surface the scene then becomes unsteady, making reasonable observation difficult and sometimes impossible. Actual appearance of turbulence can be likened to trying to look through a pane of glass while water is being poured over it. The cause is the intermingling of warm and cold air—temperature inversion. An apparent crisp clear sky as viewed with the unaided eye may well become hopeless when the telescope is brought into action, twinkling of stars or the glimmering of lights in the distance are sure signs of its presence. To measure the intensity of this interference there is in use a scale which is intended to record the severity of the condition, this scale was introduced by planetary observer Eugene Michael Antoniadi (1870-1944) of Greek birth who became a naturalized Frenchman; it bears his name. The five points of this scale are defined as:

1. Perfect seeing without a quiver in the atmosphere.
2. Slight atmospheric undulations, with moments of calm lasting several seconds.
3. Moderate seeing, with larger air tremors.
4. Poor seeing, with constant troublesome undulations.
5. Very bad seeing, scarcely allowing the making of a rough sketch.

Turbulence is not the only cause of bad seeing; thin overcast or haze may reduce clarity. To define the presence of this form of interference, or the absence of it, observers will state degree of 'transparency' in simple terms as: very good, good, fair, poor, or very poor.

SPURIOUS COLOUR

Colour which may appear to be directly associated with a lunar feature will, more often than not, prove to be an effect that has its origin with Earth's atmosphere. The condition is created either by natural causes or terrestrial contamination. Ice particles, thinly overcast sky and condensation resulting from temperature inversion can be deemed natural. For contamination, we can include volcanic dust, industrial and domestic gasses and smoke, and aircraft vapour trails.

Spurious colour will appear in two basic forms, close to the lunar terminator, shades of red will be seen to crown dark-shadowed areas, while shades of blue will be found adjacent to bright points. Regions further from the terminator and subjected to a much higher angle of illumination will display this colour in a different manner, craters will exhibit redness to the south of the feature and blueness to the north. As always, there are exceptions to this rule; the crater Plato for instance, when under high angle of illumination will completely reverse the sequence, when blue will appear to the south and red to the north. The reversal is probably attributable to the darkness of the floor of Plato.

If intensity of spurious colour is low it will be only the brightest points on the surface that will exhibit colour tinges of blue; red will be absent. The separation of the spectrum, blue-red, is caused by prismatic effect of Earth's atmosphere.

The Blink Device

To assist an observer in determining whether colour observed may be of spurious origin or not, colour filters can be introduced between eyepiece and the telescope image. I would hasten to say that there are colour outbursts that certainly do not owe their origin to spurious creations. The filters should be red and blue mounted side-by-side with no interruption between them, best contained on a rotating disk that permits rapid interchange between the colours. If the colour is positive and not of spurious origin, a reaction known as a 'blink' will take place, a lightening or darkening of a feature when seen through one filter then the other. Try experimenting with brightly coloured

ROUTINE REPORT/TLP REPORT

DATE...1980 - 7 - 23.......

NAME... A Observer............ ADDRESS. The Observatory,. Newtown,. Kent ......
....
TEL. No. 123 567...............................

INSTRUMENT USED (type & size). 12". Newtonian Reflector. x 180. x240.........

SEEING CONDITIONS. Variable. III. to. II. Transparency. fair,. some. turbulance....
(Antoniadi scale plus indication of transparency)

LOCAL CONDITIONS. Calm,. slight. mist,. Temp. 63° F.. Barom.. 30.1.................

ACCESSORIES USED. CED,. blink,. camera..
(i.e. Blink-CED- Photo-Photometer etc.)

SPURIOUS COLOUR. General. spurious. colour,. much. in. evidence. in. proximity. terminater
(please indicate extent and regions affected- alternately state if no colour seen)
****

Active from 20.30 until 21.50 UT   All east - west direction IAU

Aristarchus   20.45 UT Terminater just clearing area, west wall partially illuminated
              but interior of crater still shadowed. On the western wall, south west
              inner corner two craterlets seen also radial bands on this wall still
              contain shadow. Both eastern outer and western inner walls heavily
              tinged with spurious colour, blue. The craterlets on the west wall
              tend to merge into background when seeing falls to III.
              Beyond the terminater in the
              darkened zone, approximately
              50 miles to the north west of
              Aristarchus a high peak is   small sketch of
              illuminated by the sun.   Aristarchus

Photograph taken at 20.48 UT

20.50 to 21.10 UT Blink survey
areas scanned, Alphonsus, Theophilus, Aristarchus, Bullialdus, Plato and
Fracasterious. No response other than the permanent blink in Fracasterious
south west quadrant, usual brightening in red filter.

Gassendi   21.15 UT Much spurious colour over region, in particular intense
            redness along top of eastern rim.
            Some shadow remains along inner
            eastern wall, extends well onto   small sketch of
            floor at position between seven   Gassendi
            and ten o'clock, Shadows still
            produce much relief central peaks
            and crater floor.

21.23 to 21.45 UT CER values.
                        Piton...... 2.7
                        Pico....... 3.1   Five readings each all
                        Tycho...... 3.4   constant, no varaition
                        Bullialdus. 3.6
                        Gassendi... 3.2
                        Aristarchus
                        west wall.1.9

REGIONS SCANNED/REGIONS OBSERVED/TLP CONDITION ALERT (please indicate type of
                                                          observation)

Figure 9-3.   An example of a routine observational report. In this instance, CED, blink
device, and camera were used. Of course, these accessories will not always be available.

domestic objects; an article that is red will appear pale in the red filter and darker in the blue.

The blink device does not give a complete and final answer to the problem as unfortunately it does respond to spurious conditions on occasions, particularly if the intensity is severe. Also, a lot will depend upon colour sensitivity by individual observers. The best advice that can be given here is for an observer to experiment until he or she is aware of their own reaction in different circumstances so that when true abnormality is encountered it is easily recognizable; it will take a period of time to gain the experience.

There are certain areas on the lunar surface that will create blink reactions naturally; the southwest quadrant of Fracastorius and part of the west wall of Plato are two such regions.

TYPES OF TRANSIENT LUNAR PHENOMENA

There are four basic types of phenomena: colour abnormalities, albedo variances, obscurations, and pinpoint (starlike) flashes. All four types have been recorded with a high degree of consistency over a period of many years. The more experienced the observer the more the likelihood there is of detection of such events, for indeed there are more than a few recent sightings that, had it not been for the knowledge of the observers involved, the realization of abnormality might have easily been missed.

The causes are at present generally unknown, for consideration must be given to corpuscular radiation excitation, thermoluminescence, tidal forces, enhancement and concentration by the Earth's magnetic tail of solar plasma bombardment, UV excitation, etc. To compare or correlate each of these one would have to compare arrival of corpuscular activity at the Moon-Earth sysem, e.g., kp index and onset and occurence of magnetic storms, distance from sunrise terminator for thermoluminescence, distance from perigee and apogee for tidal, sunrise distance for UV excitation, and occurrence of magnetic storms on Earth, and Earth's magnetic tail effects. So very obviously the conclusions of causes cannot be within the bounds of amateur tasks; it is for this reason that the B.A.A. TLP network confine their activities.

Details of the types of phenomena now follow, and after the descriptions some Section case histories are included to assist the reader in understanding more about the type of situation that can be encountered.

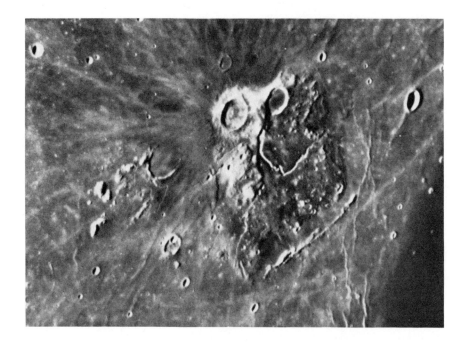

Figure 9-4.   Aristarchus, perhaps the prime TLP site. A wonderful picture, showing well the detail within the crater and, in particular, markings on the western wall. Aristarchus is approximately twenty-seven miles in diameter. The crater immediately to the right is Herodotus, and the great valley below is named 'Schroter's Valley'. (South is to the top.)
Courtesy of the Lunar and Planetary Laboratory University of Arizona.

Figure 9-5.   Plato—as a TLP site, second only to Aristarchus. Observers will find the region to be of extreme interest, changing angles of illumination creating great variations in floor shading. (South is to the top.)
Courtesy of the Lunar and Planetary Laboratory University of Arizona.

COLOUR ABNORMALITIES

Colour abnormalities are, perhaps, the most difficult of the four types of phenomena to deal with. Shades of blue, green, violet, red, orange, yellow and brown have all been reported in circumstances where the origin would not appear to be of a spurious nature. The crater Aristarchus is responsible for many and very varied colour displays; some can be attributed to the sheer brilliance of the feature against a dark background (see Figure 9-4), however, there are a large number of occurrences that defy explanation.

ALBEDO VARIANCES

Albedo variance is an unexpected brightening or dulling of a feature, a variance that does not owe allegiance to the ever-changing lunar scene. Evaluation of brightness without some form of filter aid is very difficult indeed, if not impossible. To overcome this situation the Section has introduced a filter device called the CED.

The Crater Extinction Device (CED)

The crater extinction device (CED) is a filter array designed so that the user may present any combination of filters from two series of filters positioned between the eyepiece and the telescope image. The total density of the filters can be adjusted until the feature under examination is extinguished, hence the name. The purpose of the equipment is to define albedo (brightness) values of lunar features. Neutral-density Wratten filters are used, numbers NDO 0.1 to NDO 0.9 for one-tenth stages and NDO 1 to NDO 4 or NDO 5 for the larger steps. There are two types of instruments in use. Figure 9-6 shows equipment whereby two rotating disks are employed, one in front of the other. Alternatively the unit can be constructed by placing disks side by side and enclosing them in a simple box structure. The disks are constructed of either plastic or aluminium. The two spindles that rotate the disks are twelve-way electronic switches with the wafers removed. Containing a 'click stop' mechanism, they are ideal for use because the observer can 'feel' each of the filters of different density being engaged. Tubes used for the eyepiece and drawtube will normally be produced from brass or aluminium tubing. Of the inner disks, one will carry neutral-density filters from 0.1 to 0.9, the other densities 1 to 4 or 5. In Figure 9-6 the red and blue filters for the blink device have been included and is shown as an elongated hole, remembering that it is important not to separate these colour filters. Also, each disk will have a cutout in which no filter is inserted. This enables the telescope to be used in a normal way without having to demount the CED/blink device. The diagram and detail given are not intended as constructional plans but simply as a guide to the basic concept. Any person who thinks of himself as handy probably would not experience too much difficulty in designing and making such a device.

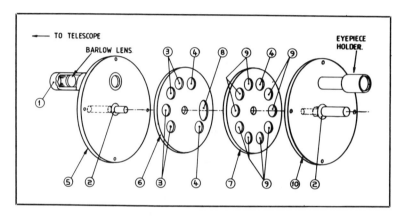

Figure 9-6. Crater extinction device (CED). The diagram depicts a typical design; no dimensions are given as it is thought that the user should construct one to suit his/her own needs. The outer discs, 5 and 10, would probably be between 15 cm and 20 cm (6 and 8 inches) in diameter. The final assembly of the unit is easily achieved by using spacers of adequate length and passing a long screw through the centre, securing it with a nut at the other end or, alternatively, spacers tapped at each end can be used. To secure a completely dust-proof unit, all that is necessary is to wrap around either plastic strip or any adhesive tape of appropriate width. If so desired, red and blue filters for 'blink' observations may be incorporated in the device.

The various parts of the device are numbered and described as follows. 1 The telescope draw-tube, usually made of brass or aluminium. 2 Click-stop mechanism. A means of rotating the two inner discs by stages will be required, and a surplus radio rotary switch is ideal; the wafer element should be removed and one is then left with a spindle which carries the mechanism. (A ten- or twelve-position switch should be obtained.) Two such switches are needed, one will be affixed to the outer disc 5 to which is fitted disc 6 and thus will rotate in stages. Likewise the same for discs 10 and 7. When assembling the inner discs to the spindle ensure that adequate clearance is maintained to avoid the filters fouling any projections; washers should be used for this purpose. 3 Cut-outs to take Kodak Wratten filters No. 96 and neutral density filters ND 1.0, ND 2.0, ND 3.0 and ND 4.0. 4 Each of the inner discs will carry blank cut-outs so that unfiltered observations may take place without removing the equipment. If 'blink' filters are fitted, one such blank position must be located next to the blink position. 5, 6, 7, and 10. These inner and outer discs must be constructed from material which is thick enough to maintain complete rigidity. Brass, aluminium, S.R.B.P., or any plastic may be used. 8 Cut-out for blink filters, red and blue Wratten Nos. W25A and 44 A. 9 Cut-outs for neutral density filters, Kodak Wratten No. 96 and Nos. ND 0.1, ND 0.2, ND 0.3, ND 0.4, ND 0.5, ND 0.6, ND 0.7, ND 0.8 and ND 0.9. *Drawing by Michael Hymer*

Using the CED

It is absolutely vital to monitor several regions during any one observational period and to take at least four measurements of each point, for it is the comparative values between regions which will give the more reliable tables. The use of comparative measures between regions eliminates variances that are caused by altitude and atmospheric conditions; at no time are single-feature direct readings of use.

ALBEDO IN EARTHSHINE

Albedo variances have been detected in the darkened portion of the Moon, the portion that is illuminated by reflected light from Earth (Earthshine). The method of observing the darkened area of the Moon is to remove the area illuminated by the Sun from the field of view and allow the eye to adjust for a few minutes. You will be surprised at the amount of detail that can be distinguished, certainly within the lunar phases up to first quarter and again from last quarter. Any abnormal brilliance, or even colour, will be easily detected. Again Aristarchus has been responsible for events in this condition.

OBSCURATIONS

Obscuration is a form of TLP whereby a part of a formation exhibits haziness or opaqueness while immediate surroundings remain sharp. Here the skill of the observer is paramount. Light and dark points on the surface can and will merge at the limit of resolution of the telescope. This must not be confused with true obscuration. A thorough knowledge of an area under surveillance at all stages of the lunation is a must, particularly when observing rugged terrain.

STARLIKE FLASHES

Starlike flashes are the last of the four known types of TLP, brilliant pinpoints of light flashing out for split seconds against the lunar background. Recent reports indicate that it is usual for this phenomenon to repeat several times in a specific region within a short space of time. Known sites are the locality between Plato and Mt. Pico, Grimaldi, and in and around Aristarchus. Care must be taken not to confuse this form of abnormality with pollen contamination in our atmosphere, which will be seen as gently sparkling points as opposed to the sheer and stark brilliance of a starlike flash.

KNOWN TLP REGIONS

Many areas have been the subjects of TLP reports. The list below details just a few. (Aristarchus and Plato have been responsible for a high percentage of reports by the Section in recent times.)

Alphonsus	Daniell	Menelaus	Torricelli 'B'
Aristarchus	Gassendi	Mt. Pico	Theophilus
Bullialdus	Grimaldi	Plato	Tycho
Cape Laplace	Manilius	Spitzbergen Mts.	

TLP CASE HISTORIES

From the Section records, some brief extracts of recorded events are now given. It is interesting to note that in not a few instances two or more types of phenomena have been present at the same time. More than 200 abnormal appearances were recorded by the Section in the period from 1978 December to 1983 December.

1978 April 20th (20.30 to 22.35 hours UT).
Cape Laplace –yellowish brown colouration toward tip of peninsula, response in blink device. (Observers engaged: G.W. Amery & P.W. Foley.)

1980 April 18th (20.00 to 21.56 hours UT).
Aristarchus in Earthshine –at 20.00 hours was dull, barely discernable. 20.16 hours interior of crater suddenly displayed flashes, likened in appearance to St. Elmo's fire, seemed to originate inner southeast (IAU) corner and spread rapidly to illuminate entire interior. By 20.28 hours brilliance had subsided but blue incandescence interior and exterior, spreading to Herodotus remained. At 21.07 starlike flashes were observed toward inner southeast (IAU) corner. Between 21.07 to 21.56 hours UT variable brilliance was detected. Other regions monitored in Earthshine were stable and unaffected by colour or albedo changes. (Observers engaged: G.W. Amery, A. Cook, J.D. Cook, P.W. Foley, P. Madej, P.A. Moore, F.W. Peters, M. Price, G.H. Ricketts, J. Saxton.)

1980 September 24th-25th (20.48 to 02.55 hours UT).
Plato region –two billiant starlike flashes of split second duration. The first was seen just north of Mt. Pico while the other occurrence was just south of the southernmost extremity of Plato. As for Plato itself a loss of detail (obscuration) was detected on and over part of the northwest (IAU) wall. Blink reaction was obtained from floor, in blue filter central craterlet was seen, and from it

issued dark radial streaks, one to the south, another to the southeast. (Observers engaged: G. Blair, J. Cook, M. Cook, P.W. Foley, P.A. Moore, J. Pedler, J.H. Robinson.)

1983 January 29th (20.35 to 23.15 hours UT).
Torricelli 'B'–whilst taking routine CED measures reached Censorinus, immediately became aware that lying to the southwest (IAU) was an unbelievably bright point, craterlet Torricelli 'B'. The albedo value went beyond maximum for the CED device, the observer could not extinguish the feature at a value of 5.5 and concluded that the true value was over 6.0. Other readings for that night were, Aristarchus 3.8 and Censorinus 3.5. Visually, Torricelli 'B' was seen to have an incredible halo around the inner rim. Another observer described the colour like that seen in ultra violet sterilising lamps. By 22.40 hours UT the brilliance subsided and the colour underwent a change to deep rose-purple. (Observers engaged: G.W. Amery, B.W. Chapman, J. Cook, M. Cook, P.W. Foley, M. Mobberley, P.A. Moore, G. North, F.W. Peters.)

From these examples it will be seen that a TLP can exist as a single and simple colour event (or for that matter it may just be an obscuration or albedo change) or as a quite complex combination of two or more different types of phenomena.

Please note that for American readers, the term TLP becomes LTP (Lunar Transient Phenomena).

RECURRENT LUNAR PHENOMENA (RLP)

There is a final aspect to consider, visual oddities that owe their origin to angle of illumination or phase of the Moon. For instance, the southernmost point of the northern peak of the central mountains of Gassendi exhibits deep red colouration during the period of approximately 18 to 24 hours after the terminator has receded. Cape Laplace will display a dark shadow at the western (IAU) tip of its peninsula long after all other shadows in the region have been obliterated by high angle of illumination. The crater Bullialdus has a feature on the west wall which casts a spectacular round shadow while the crater is still under relatively high angle of illumination. Atlas, that has a depression in the inner southwest corner, throws an immensely black shadow just after local noon. There are uncounted traps for the unwary such as these, and for this reason the Section is listing these appearances as they are identified. For, perhaps, want of a better name the author decided to refer to them as Recurrent Lunar Phenomena (RLP), though of course, they are not true phenomena.

Whom to contact:

The co-ordinator of the United Kingdom team is the author:

P.W. Foley, Tree Trunks, Nettlestead
Maidstone, Kent, England. ME18 5HJ

Within the United States there is a similar organization, and the contact is:

Winifred S. Cameron
200 Rojo Drive
Sedona, Arizona 86336

Acknowledgements
The author would like to express his appreciation of the help given by two good friends, Mrs Winifred S. Cameron for help on some technical aspects of the chapter and Ewen Whitaker of the Lunar and Planetary Laboratory, University of Arizona, who has permitted the use of the splendid photographs.

ABOUT THE AUTHOR
With a long-standing interest in astronomy and in the Moon in particular, Peter Foley has made many observations of the lunar surface. A chance observation of an anomalous appearance of the formation Prinz in 1974 led to an absorbing interest in transient lunar phenomena, a field in which he now specialises. He uses a 295 mm reflector at his observatory in Kent, is currently Assistant Director of the Lunar Section and co-ordinator of the Section's TLP project. He has made many valuable contributions to the study of these controversial phenomena.

10

THE OCCULTATION PROGRAMME
by A.E. Wells

The timing of lunar occultations of stars is now one of the Lunar Section's major programmes. It provides valuable data for studies of the lunar motion and of the topography of certain limb regions. It is a field in which amateurs can make signficant contributions to current research, but it is only fair to stress, at the outset, that while the work is challenging, it often is frustrating, considerable tenacity being required for success. A single observation is of little value; what is required is a steady contribution over a long period of time. Only simple equipment is required, however; indeed, a small portable telescope can often be more useful than a large, permanently mounted one. It goes without saying that occultation observers must adopt a scientific and objective approach to their work and must ensure that their observations are entered on the standard report forms, which can be obtained from the International Lunar Occultation Center from the International Occultation Timing Association (IOTA), or, for members of the B.A.A. Lunar Section, from the Section Co-ordinator.

An occultation is defined as the phenomenon occurring when one celestial body is concealed by another. It should be distinguished from an eclipse, which is the passage of a non-luminous body into the shadow of another. (Strictly speaking, a solar eclipse is really an occultation). The Lunar Section is primarily concerned with lunar events that occur as a result of the easterly motion of the Moon across the sky. The term 'occultation' includes both disappearances (immersions) and reappearances (emersions) and, although these occur at both bright and dark limbs, they are normally observed only at the latter because bright-limb glare irradiation reduces the accuracy of the observation. Thus, immersions are usually observed during the waxing phases and emersions during the waning phases of the Moon, the latter often requiring work in the small hours! Anyone able to work at these times can make an especially valuable contribution by timing emersions.

Essentially, the observer is asked to record the time of occultation to a precision of one tenth of a second, and to supply his position in latitude and longitude to the nearest second of arc and his height above sea level to the nearest 30 metres (100 feet). This information can usually be obtained from U.S. Geological Survey Topographic Maps, or in the U.K., from Ordnance Survey maps at public libraries. In the U.K. the Section Co-ordinator will convert Ordnance Survey map references into latitude and longitude. A U.S. Geological Survey map 7'.5 (seven-and-one-half-minute) series will be required to obtain the necessary accuracy. Such maps can also be obtained from U.S. Geological Survey.

First, it is necessary to know approximately when and where an occultation will take place. The main source of predictions is from the U.S. Naval Observatory, Washington, D.C., which generates predictions for individual observers, in some cases for stars down to 10th magnitude, the need depending on the equipment owned by the observer. Predictions in the Lunar Section *Circulars* extend down to mag. 9.0, in the B.A.A. *Handbook* down to 7.5, and in the Royal Astronomical Society of Canada (R.A.S.C.) *Observer's Handbook* down to 6.0. These predictions tabulate the necessary information in standard form, much of which is self-explanatory. The number of the star in the *Robertson Zodiacal Catalogue* (ZC), the *Smithsonian Astrophysical Observatory Star Catalog* (SAO), the *Bonner Durchmusterung* (BD), or the United States Naval Observatory (USNO) catalogue, must be quoted in the observer's report to enable the star to be identified. The column headed 'Ph' indicates whether the event is a disappearance (D) or a reappearance (R). The time of occultation at the standard station for which the prediction is issued is given in U.T. When an observer's station is a considerable distance from the standard station a correction must be applied, and the *a* and *b* coefficients enable this to be done.

One final piece of information is required before the star can be identified. This is the position angle of the event, i.e., the position of the point of occultation on the lunar limb. Predictions given in the U.S.N.O. *Predictions* show the cusp angle, which is simply the angular distance between the point of occultation and the nearest cusp, supplemented by the letters 'N' or 'S'. In the U.K., the equivalent predictions are also published in the Lunar Section *Circulars*. It must be borne in mind, however, that cusp angle will change with distance from the standard station. Additional information given in the *Predictions* includes the star number in the *Smithsonian Astrophysical Observatory Star Catalogue* (SAO), its RA, its declination, its spectral class, and also a code indicating if it is a binary.

The timing of occultations is of vital importance. A precision of 0.1 s is demanded, but this is not as difficult as it may sound, provided one of the accepted techniques is adopted. The majority of observers use a stopwatch and a primary time standard such as the telephone time signal or a radio time signal such as that transmitted by WWV and CHU or in the U.K., from Rugby (call sign MSF) or from Nauen, Germany (call sign DIZ) or, in the U.K., the telephone time signal. (In most cities in North America telephone time is not accurate enough; exceptions are 303, 499-7111, WWV, Fort Collins, Colo, and 202, 653-1800 [U.S. Naval Observatory master clock].) Alternative techniques include the eye-and-ear method (whereby the observer listens to a continuous time signal and estimates the fraction of a second between beats), the clock-and-camera method, and others.

Using the stopwatch/time-signal method, the observer sets up his telescope at least 30 minutes before the event, allowing plenty of time to locate the star. A low-power eyepiece is usually best except when the star is faint or there is a glare from the Moon, in which case it is sometimes advantageous to use a medium-power. High magnification is not required. The telescope is set to track the star, and the stopwatch, fully wound and zeroed, is held in the hand. Where a slow motion is fitted, it must not interfere with the observer's comfort in also holding the watch! Careful concentration on the star is required until the moment of extinction (which is instantaneous, invariably taking beginners by surprise). The stopwatch button is pressed simultaneously with the observed extinction of the star. As soon as possible afterwards, a time signal is obtained; tenths of a second are determined by examining the stopwatch while it is running and listening to the seconds' beats. The watch is stopped on a suitable minute or known second. This time is noted and the stopwatch reading subtracted from it, to give the 'raw time' of occultation.

No matter how quickly an observer presses the stopwatch button, a brief time elapses beween observation and response. This is a physiological phenomenon and is called the 'personal equation'. It is usually between 0.3 s and 0.5 s, depending on the observer's alertness and sky conditions. Various methods are used by experienced observers to determine their personal equations, but it is best that beginners do not attempt this. Nevertheless, they should develop an ability to estimate time and say whether they think that their reaction was slower than usual, as often happens in the case of faint stars.

Further coded information is required to enable the International Lunar Occultation Center to weight the observation, such as an estimate of observing conditions and of the accuracy of the timing. The observer is in the best position to supply this information. All the required data, including the star number and any appropriate remarks, should be recorded on the standard form and returned to the International Lunar Occultation Center within six months, or, in the U.K. to the B.A.A. Lunar Section Co-ordinator.

THE OBSERVATION OF GRAZING OCCULTATIONS

In the light of a distant star, the Moon casts a full-size shadow nearly 3500 km (2200 miles) wide, upon the Earth. As this shadow, distorted only by the Earth's curvature, passes rapidly across its surface, most observers within the shadow see the total occultation described previously in this chapter. Observers situated within a narrow zone at the north or south edge of the shadow, however, may observe a number of disappearances and reappearances within a few minutes, as the star passes behind mountains and valleys on the lunar limb. A grazing occultation can be spectacular indeed, and a successful observation of such an event provides very valuable data for studies of the Moon's position in declination and of the profile of the limb. In general, only observers with small, portable telescopes will be in a position to carry out the necessary observations of grazing occultations.

Obviously, predictions for grazing occultations cannot be provided in the same format as those for total events, because the observer needs to know not only the time, but also the place to which he must go to obtain the maximum information. Observers within travelling distance of any of these tracks (the distance an individual is prepared to travel is a personal decision) should request a set of predictions directly from the I.O.T.A. In the U.K., notification is given in the B.A.A. *Handbook* in the form of tracks across an outline map and, from time to time, in the Lunar Section *Circulars*. Similar maps for North America are published in the R.A.S.C. *Observer's Handbook*. In the U.K. these will consist essentially of two sets of co-ordinates which, if plotted on an Ordinance Survey map, yield a pair of almost-parallel lines, denoting the graze limits. Thus, for a northern limit graze, observers south of these lines would see a brief total occultation, while those north of the lines would see a near miss. Only in the narrow track between the lines can the spectacular events of a graze be expected. In the United States the line which appears on the map is different; it indicates the centre point, i.e., the optimum place for observation. Other data in the predictions include times of mid-graze for all the positions given, a height coefficient, and information on the star and its position. When a Watts profile chart is available, its use will enable observers to position themselves to even greater advantage.

Ideally, a graze is best monitored by a team of observers working under the direction of an experienced leader and using portable telescopes set up at intervals along a line extending diagonally across the track. Once the observing site has been chosen, the team leader, at least, should survey it in daylight, making plans to position his observers where the sky is unobstructed and, using information obtained from the Watts profile chart, where most events are likely to be seen. If there are sufficient observers, at least one should be stationed beyond each track limit in case a small error has been made, either in the predictions or the planning.

Timing the events observed during a graze is best done with a tape recorder on which is recorded a continuous time signal such as the transmission WWV, CHU, or Rugby in the U.K. and DIZ in Germany. The observer records his comments on the tape in the form of a sharp-toned 'In' or 'Out', according to the event. Some observers prefer to use a mechanical 'clicker' instead of vocal comments. If no radio signal is available, a time signal from a nearby telephone call-box can be recorded immediately before and after the graze, the tape recorder being kept running and held in the same position throughout the observation.

Whichever method is used, a permanent record of the observation results, which can be played back at leisure, timing the intervals between the marker signals and each event with a stopwatch. Always ensure that new batteries are used in the tape recorder throughout. Each observer should record his own observations on the standard report form, entering the co-ordinates and height of his particular observing station also. The team leader is responsible for seeing that this is done and for collecting the report forms and forwarding them to the International Lunar Occultation Center without delay. A copy of the reduced observations will be sent to him in due course.

"PROJECT FADE"

The occultation of a star by the Moon's limb is instantaneous, due to the absence of any appreciable lunar atmosphere. Occasionally, however, a star will seem to fade into extinction over a very short, but noticeable, period. Several explanations for such anomalous behaviour have been advanced, one of the most common being that it is a binary star. The light from one component becomes extinct a few milliseconds before that of the other. In the case of near-grazes, the lunar limb may split a double star so that one of the stars never becomes extinct. A gradual extinction of the star's light is called a "fade".

It has also been suggested that, when the star passes behind a lunar slope at a critical angle, diffraction of its light may contribute to the effect. There remains the possibility, of course, that at least some fading occultations may be due to purely local causes such as atmospheric conditions.

The Section's "Project Fade" was initiated in 1972 by Patrick Moore (who was then Director), after he had observed a fading occultation during a Pleiades passage of the Moon. The aim of the project is to maintain a complete record of every anomalous occultation, for future analysis. Copies of reports are also sent to Dr David Dunham in the United States, who is compiling a double-star catalogue.

All observers are therefore asked to report any anomalous occultation behaviour to the Co-ordinator immediately, giving full details of location, instruments, conditions, etc. All reports will be acknowledged and published in the Lunar Section *Circulars*.

EQUIPMENT FOR OCCULTATION OBSERVING

Beginners often ask what sort of telescope is best for occultation work, but this is not a question which can be answered simply, because so much depends on sky conditions and the phase of the Moon. When the sky is dark and clear, and the Moon is in the crescent phases, small telescopes will easily show stars down to 7th or 8th magnitude. Even a pair of good binoculars can be put to use, provided they are rigidly mounted. On the other hand, if the sky is slightly hazy and the Moon is gibbous, there is often so much scattered light that it can be difficult to see even a 6th magnitude star with a large instrument! Another distinct advantage of a small telescope is its portability, which is important when 'tree dodging' becomes necessary or when doing graze work. So do not despair if you have only a small telescope or binoculars. Just ensure that you have a sturdy mount with smooth slow motions—and then look for opportunities during the crescent phases and when the sky is especially clear.

A stopwatch, too, is almost indispensable. The best for this purpose is one with a ten-second sweep, calibrated in tenths of a second. It is worth paying a little extra to get a good quality watch, and some observers may like to consider a split-action watch, which can be useful when two stars are occulted within a short time-period. This circumstance does not often occur, however, and it may not be possible to justify the extra expense. (But if you do get one, be sure it is reliable and accurate, otherwise use a tape recorder, or the 'eye and ear' method, or even two stopwatches for binary star occultations.) The watch should be fully wound and held in the hand for some while before the event. It should be held in the same position while running and should be frequently rated against a standard time signal.

When the telephone time signal cannot be used, a shortwave radio capable of receiving the transmission by WWV (on 2.5, 5.0, and 10.0 MHz) or CHU (on 3.35, 7.35, and 14.65 MHz) is invaluable. In the U.K. the transmission from Rugby, call sign MSF (on 60 kHz) should be used. Before deciding to purchase such a receiver, observers are advised to contact the Co-ordinator, who is always ready to offer advice.

Finally, a simple cassette tape-recorder, although not a necessity for routine work, will be found useful if graze work is contemplated.

FURTHER READING

The following sources should be consulted for further information on occultation observing:

Lunar Section *Circulars*

Handbook of the British Astronomical Association

H. Povenmire, *Graze Observer's Handbook,* Vantage Press, New York.

IMPORTANT ADDRESSES

International Lunar Occultation Center, Astronomical Division, Hydrographic Department, Tsukiji-5, Chuo-ku, Tokyo, 104 Japan.

Washington Distribution Section, United States Geological Survey, 1200 South Eads Street, Arlington, Virginia 22202, U.S.A.

United States Naval Observatory, Washington, D.C. 20390, U.S.A.

International Occultation Timing Association

U.S.: P.O. Box 596, Tinley Park, Il 60477, U.S.A.

Europe: Hans J. Bode, Bartold Knaust Str. 8, 3000 Hanover 91, West Germany.

ABOUT THE AUTHOR

A member of the B.A.A. Lunar Section for many years, Alan Wells' special interest is the observation and timing of lunar occultations. An electronics engineer by profession, he has designed and constructed various devices to improve the timing accuracy for these phenomena. He is a member of the Lunar Section advisory committee and serves as Lunar Occultation Co-ordinator for the Association.

11

THE STUDY OF LUNAR TOPOGRAPHY
by D.G. Buczynski

After more than 350 years of intense study, we have a near-complete knowledge of the topography of the near-side lunar surface down to metre size. The vast amount of photographic data available is staggering, and the history of the formation of the lunar surface can now be unravelled from it. Apart from an area around the south pole, lunar cartography is complete, and there remains little for the amateur to achieve using the classical methods of lunar observation.

THE PROSPECT FOR POST-APOLLO VISUAL LUNAR WORK

The Moon nevertheless attracts many amateur astronomers to direct their telescopes towards her, and the view obtained even from a small aperture instrument is quite astounding. While it is probably true that there is little likelihood of any amateur work today resulting in a major revision of existing lunar maps, there does remain a programme in which the amateur can participate and contribute to our knowledge of the topography of the lunar surface as seen from the Earth.

Simply put, the prospects are these. During the past 350 years of lunar mapping, the emphasis of study has been directed towards positional accuracy and the filling in of fine detail upon lunar maps. To obtain these data, observers would study the numerous features under all conditions of illumination. Grazing and low illumination angles enable small-scale features to be seen, although such features disappear under higher illumination angles. The resulting changing face of the Moon as seen from the Earth, with respect to individual features, has never been properly chronicled. There is little in the literature that describes the changing 28-day scenery of a lunar feature (the time that a particular feature is illuminated is actually about 14¾ days). Information such as the timing of the first appearance or illumination of crater walls, peaks and floors, and of ridges, rilles, and other lunar features during a lunation is worthy of publication. The systematic recording of these regular changes as seen through Earth-based telescopes would prove a challenging observational effort for a team of amateur astronomers to become involved in on a long-term basis.

ILLUMINATION FACTS

Quite obviously, from our point of view, the cause of the varying aspects we see is the changing illumination of the lunar surface by the Sun. Most basic books on astronomy will graphically describe the rotation of the Moon and Earth in their orbits around the Sun, these being the factors involved in our seeing the lunar phases. There are, however, some further fine details to be added to this explanation which affect the illumination conditions of lunar features as seen from Earth.

The terms used to define when a particular feature is illuminated are the Sun's selenographic colongitude or the lunar phase angle. Selenographic colongitude is the most commonly used figure, although the phase angle does help us visualize from which direction the illumination is coming. When the Moon is in conjunction, i.e., between the Sun and the Earth, the sub-solar point on the Moon is almost opposite the sub-terrestrial point, i.e., they are 180° apart in longitude on the assumed lunar sphere. As the Moon shifts in its orbit around the Earth, the sub-solar point moves along, or near to, the Moon's equator at a rate of about 0°.5 per hour, or 12°.2 per day towards the west (IAU). The sub-terrestrial point would remain fixed centrally on the Moon's disk were it not for the two principal librations (it does define the shifting centre of the Moon's disk for the terrestrial observer.) The westward motion of the sub-solar point results in a decrease of angle at the Moon's centre defined by the points G, E' and S (see Figure 11-1) between New and Full Moon and an increase thereafter. This varying angle is the phase angle. At New

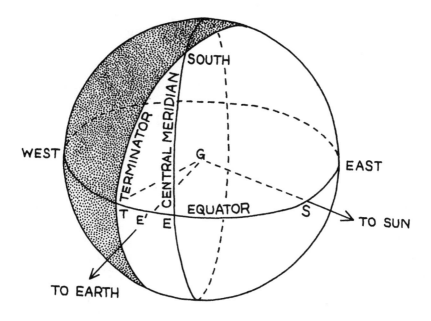

Figure 11-1.   E is the mean position of the sub-terrestrial point, T is the position of the terminator on the equator, angle SGE′ is the phase angle, and angle EGT is the Sun's selenographic colongitude.

Moon, the phase angle is near -180°; at First Quarter, when the morning terminator passes through the sub-terrestrial point, it is -90°; at Full Moon, when the sub-terrestrial and sub-solar points are at the same longitude, the phase angle is near 0°; and at Last Quarter, when the evening terminator passes through the sub-terrestrial point, the phase angle is +90°. Finally, when the evening terminator falls on the western limb, the phase angle is near -108°, whereupon the cycle begins again. When using the term selenographic colongitude, it is well to bear in mind the phase angle, as this will immediately indicate how obliquely the solar rays are illuminating the feature under study.

The complications of libration, in both latitude and longitude, do not affect the timing of the first illumination of a feature in relation to its selenographic colongitude. Sunrise on a feature always occurs at the same selenographic colongitude. Libration is only a line-of-sight effect directed at the lunar surface from Earth. A factor which does affect the first illumination of a particular feature is that during a year the position of the Sun varies by 1½° both north and south from the plane of the Moon's equator. This affects the altitude of the Sun above the horizon of a particular lunar feature and this differs from lunation to lunation. Consequently, the illumination may strike a feature at different angles even though the selenographic colongitude is constant. This effect should be borne in mind when we attempt to chronicle the sequential changes observed during a lunation and compare them with other sequences obtained at different lunations by either the same or other observers. A more complete measurement of the illumination other than the selenographic colongitude is the measure of the Sun's selenographic latitude; both of these values give us a complete measure of solar illumination as seen from the Sun. For an exact measure of solar illumination of a feature as seen from Earth, we use the feature's known latitude and longitude on the lunar globe combined with the Sun's azimuth and elevation in the lunar sky of that feature.

The effect libration has on the view we obtain of lunar features depends on their position upon the lunar disk; features near the disk centre are little affected, while features situated near the lunar limb are grossly affected and can become unrecognizable or even invisible. The result of libration on a particular feature at identical colongitudes is a foreshortening effect of the type that changes a circular crater into an apparently elliptical one, and it changes the lengths of the shadows, integrated intensity, and contrast markings. This degree of foreshortening is demonstrated in a table prepared by W. Haas*.

EFFECT OF LIBRATION

Longitude	0°	15°	30°	45°	60°	75°	90°
Average	1.000	.966	.866	.707	.500	.259	.000
One maximum	.993	.990	.926	.796	.612	.387	.135
Another maximum	.993	.929	.791	.605	.379	.126	

Latitude	0°	15°	30°	45°	60°	75°	90°
Average	1.000	.966	.866	.707	.500	.259	.000
One maximum	.993	.990	.919	.786	.599	.371	.118
Another maximum	.993	.929	.801	.619	.395	.144	

---

* From Haas, W., "Does Anything Happen on the Moon?", *Journal of the Royal Astronomical Society, Canada,* vol. 36, page 22 (1942).

LUNATION SEQUENCES

The varying aspect we see of lunar features during a single lunation is a very striking and complex affair. The delineation of these changes in the form of drawings and detailed notes is a challenge to the observational skill of all astronomers. However, once begun, the completion of a detailed sequence becomes very special to the observer, and much information and many points of interest, some possibly even with controversial aspects, may emerge. Observers should follow a planned programme when embarking upon this type of observation. Haphazard or carelessly made observations will not lead to good quality results. It would be well to attempt sequences for only perhaps two formations in a year and to pursue these with all vigour until satisfactory observations are made. It is best to try and obtain sequences that can be completed in one lunation. A good sequence for one lunation would comprise about six or eight drawings and extensive notes and timings. It may not be possible to obtain the sequences all in one lunation, in which case the observer should try to select lunations with complementary librations. A general sequence would comprise two or three drawings at sunrise, one or two between sunrise and local noon, another one or two between noon and sunset, and two or three at sunset. Of course these observations may be extended to become as numerous or as detailed as the observer wishes, but it is important to try and have a balanced sequence, e.g., not having many observations up to local noon and none after.

The observational procedure is as follows. Make observations of the formations covering the changing appearance throughout a complete diurnal passage of the Sun across them. Drawings of these formations made at any time during a lunation should be prepared on a 100 mm by 125 mm (4 x 5-inches) white card to a generous scale, e.g., 3.25 metres (130 inches) to the lunar diameter is a large scale. (It is not necessary to adopt a particular scale for all drawings; it is better to adopt a scale that suits the lunar formation being studied and the atmospheric seeing conditions at the time of observation). Try not to attempt too large an area for study as this makes positional work and subsequent identification difficult. Drawings should be accompanied by the following data printed on the reverse face of the card.

Instrumentation. E.g., 235 mm f/8 reflector (put Cass for Cassegrain, O.G. for refractor, Cat for catadioptric, Mak for Maksutov, etc.)

Ocular and magnification. E.g., 12.5 mm Orthoscopic x 144.

Seeing on the Antoniadi scale. E.g., seeing II-III.

Conditions adversely affecting observation. These include haze, twilight, fog, artificial light, wind, etc.

Object under study. E.g., Piccolomini.

Time of commencement and completion of observation. E.g., 1983 August 31d 02h 30m UT to 04h 30m UT.

Lunation number. E.g., lunation 751.

Sun's selenographic colongitude. E.g., 153°.7 (mid-observation).

Indication of the direction of the cardinal points. I.e., north, south, east, and west (IAU sense).

Detailed observational notes should accompany sequences, these having been made during the observations and left uncorrected afterwards. References to interesting points should make the text of any observational notes: e.g., when a crater wall is first lit or is covered with shadow, any isolated bright or dark features within the formation, any details of colour phenomena, intensity estimates of variable contrast markings, etc. High quality photographs with high resolution taken *during* the observational period would make a valuable addition to the record.

FEATURE MORPHOLOGY

Taken on their own, lunation sequences are limited in the amount of information that they can provide. However, the information returned by the Lunar Orbiter and Apollo missions is readily available and this, used in conjunction with telescopic data such as lunation sequences and good-quality Earth-based photographs, can provide us with sufficient information to attempt feature morphologies. Many good photographic atlases have been produced in the U.S.A. during the last 20 years. Dimensions, positions, and feature profiles also are available for such a study. We could describe a formation in its fullest terms as seen from Earth using all these data giving such items as the appearance of the feature at approximate solar colongitudes, the scale of variation of the intensity of contrast markings, details of approximate shadow extent with solar colongitude, and permanent colour sites plotted against solar colongitude. In short, the successful undertaking of this type of programme would provide an observational reference for lunar observers on a scale that has not been attempted before.

CONCLUSION

The Moon will always be an attractive object for telescopic study by amateur astronomers. There is always some part of any month during the year when it is well placed for observation. The detailed nature of the view afforded by a small telescope is quite astounding, and the spectacle during an observational session blessed with good seeing is among the very best views that Nature can provide. In most active pastimes, participants usually want to progress from learning the basics to being a master in their chosen activity. So it is with amateur astronomers, and the Moon is an excellent place to begin, just for its sheer beauty. Progressing to the undertaking of a regular observational programme designed to increase our overall understanding and awareness of the lunar surface is to be well on the way to becoming a master of one's pastime. It will, if executed in a carefully planned and exacting manner, also leave a useful legacy for future generations of lunar observers. The above outline indicates a comprehensive long-term programme to which the amateur astronomer could contribute with satisfaction.

ABOUT THE AUTHOR

Member of the B.A.A. for many years and currently a member of the Lunar Section advisory committee, Denis Buczynski's special interest is the preservation of the classical methods of lunar study. He has made many valuable observations of the Moon, himself, and has revived this interest among Lunar Section observers. He is currently engaged in collating sequence drawings of selected features. He successfully launched, and is now editor of, the Section's topographical publication *The New Moon*.

12

LUNAR PHOTOMETRIC AND COLORIMETRIC PROPERTIES
by E.A. Whitaker

Now that observers are paying more attention to the search for possible changes or anomalies in brightness, albedo, and colour on the lunar surface, it is desirable that they know some of the basic facts concerning these properties, and of the implications the properties have for the physics and chemistry of the surface layer.

PHOTOMETRIC PROPERTIES

Although it was anciently thought that the Moon might be a mirror, reflecting the Earth's surface land and sea masses and clouds, we now know, of course, that it *scatters* incident light, i.e., parallel rays of sunlight are reflected in all directions after impinging upon the materials of the lunar surface. It might be thought that this surface acts similarly to most natural terrestrial surfaces of a generally matte nature such as soil, gravel, sand, or even barren, weathered rock faces, but this has long been known not to be the case.

A truly matte surface scatters incident light isotropically, i.e., with equal intensity in all directions above that surface, and is known as a *Lambert surface* if 100% of the incident light is scattered back into space. In what follows, this will be referred to as a *white Lambert surface*, whereas a *Lambert surface* will signify an isotropic scatterer that absorbs some of the incident light. A plane surface of this kind, when uniformly illuminated by parallel light at some given angle of incidence, gives a constant reading on a spot photometer, no matter from what angle or distance the reading is made, provided, of course, that the surface is angularly larger than the spot.

For a Lambert sphere illuminated by sunlight (essentially parallel), the surface brightness is simply proportional to the cosine of the angle of incidence; hence the sub-solar point (cosine 0) has the greatest brightness, diminishing to zero at the terminator (cosine $90°$), and isophotes are small circles centred on the sub-solar point. At Full phase, the terminator falls on the limb, and the limb darkening is maximum. The integrated light at the Quarter phase is $1/\pi$ (i.e., 0.318) of that at Full, and the phase curve is smooth across the Full phase point (see Figures 12-1, 12-2, 12-3, and 12-4).

The lunar sphere differs from a Lambert surface in several ways. In addition to the permanent difference in reflectivity (albedo), the lunar sphere differs as follows:

1. The sub-solar point is not the brightest spot.
2. Isophotes are illumination meridians; thus the limb is equally bright from pole to pole—it is also the brightest part of the disk.
3. The Full Moon is equally bright everywhere; i.e., there is no limb darkening.
4. The integrated light at the Quarter phase is about 0.08 of that at Full phase.
5. The phase curve is sharply peaked across the Full phase point (see Figures 12-5 and 12-6).

Figure 12-1.   Angles of incidence (i), emergence (e) and phase (g) illustrated. I and E are incident and emergent rays respectively, and PN is normal to the scattering surface at P.

Figure 12-2.   The apparent brightness of a Lambert surface, which is independent of viewing angle or distance, depends only upon the illumination intensity. Thus brightness in area B (angle of incidence $= i$) equals $cos\ i$ times brightness in area A (normal incidence).

Figure 12-3.   Isophotes for four chosen brightnesses (normalized to 1.0 for zero phase brightness) on a Lambert sphere, as viewed at phase angles of (a) $90°$, (b) $45°$ and (c) $0°$.

Figure 12-4.   Graph of the integrated light (L) of a Lambert sphere with varying phase (g), normalized to 1.0 for zero phase angle.

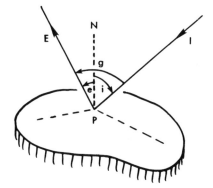

Figure 12-1.

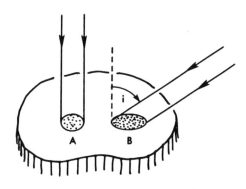

Figure 12-2.

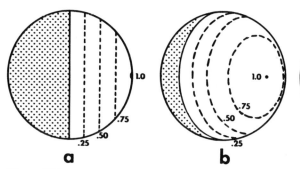

a   **b**   **c**

Figure 12-3.

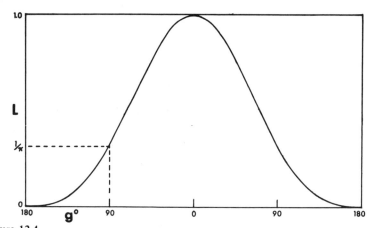

Figure 12-4.

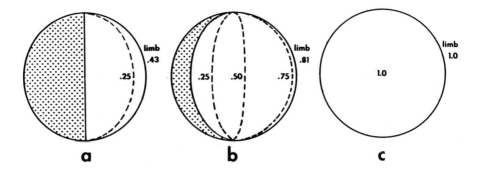

Figure 12-5. Isophotes for same four chosen brightnesses as in Figure 12-3, normalized to 1.0 for zero phase angle brightness (but excluding the opposition effect), on a constant albedo Moon. Phase angles as in Figure 12-3.

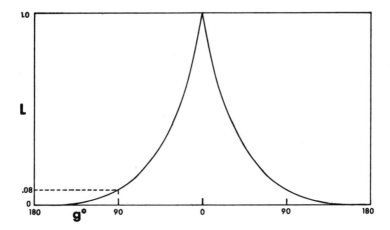

Figure 12-6. Graph of the integrated light (L) of the Moon with varying phase angle (g), normalized to 1.0 for the 'fictitious Full Moon' (i.e., excluding the opposition effect). The integrated light at zero phase is 1.5 times greater than that of a Lambert sphere of the same size and geometric albedo.

Past attempts to explain the lunar observations in terms of a Lambert surface having various amounts of topography and unresolved roughness of a generally similar nature (i.e., craters, mounds, ridges, rilles, peaks, etc.) all failed dismally. This is largely because, for craters, the brightly illuminated sunward slopes are approximately compensated for by the poorly illuminated or totally shadowed opposite walls. A similar situation holds for positive relief features, so that the overall effect on the photometric properties is minimal. The greatest divergence occurs at locations near the terminator, where long shadows overcompensate for the bright, sunward slopes. Taking this topography criterion to its limits, it was found that even highly vesiculated ('aa') lava surfaces, deep cracks, etc., could not duplicate the lunar observations.

It is now known that the non-Lambertian photometric properties of the lunar surface are due to the highly porous nature of the outermost few millimetres of the regolith (the ubiquitous, fine-grained soil). The essential property is that sunlight can filter down and illuminate sub-surface grains, the scattered light from which can filter back out again through gaps in any direction. Thus at the Full phase, all grains cover their own shadows, and the Moon's limb is as bright as any other location on the disk. The same effect can often be observed from an aircraft when its shadow falls on trees or other vegetation; the shadow (or its location) is seen to be surrounded by a bright halo.

The brightness of such a porous, plane surface can be expressed mathematically, assuming parallel incident light and an isotropic but random array of spheres (representing the regolith grains):

let b_o = brightness of surface at phase angle zero (as measured with a spot photometer)

b_g = brightness of same surface at phase angle g

i = angle of incidence of light

e = angle of emergence of light (see Figure 12-1 for illustration of these angles)

then $b_g/b_o = k_g \left(\dfrac{\cos i}{\cos i + \cos e} \right)$

except where both i and e approach $90°$ and are unequal. When $i = e = 90°$, the term in brackets may be taken as being 0.5.

The factor k_g depends on the porosity of the surface, and also on the average phase function of the spheres plus secondary scattering. By adjusting these parameters properly, an excellent fit to the lunar photometric observations can be obtained; the mean bulk density of the surface layer calculates at 38%, i.e., 38% solid grains, 62% space, and the grains approximate to Lambert scatterers (but with some extra scattering in the general direction of the incident light). The factor k_g is tabulated herewith at 5° intervals:

$g°$	k_g	$g°$	k_g	$g°$	k_g
0	2.00	50	0.75	100	0.37
5	1.75	55	0.70	105	0.35
10	1.55	60	0.65	110	0.33
15	1.39	65	0.61	115	0.31
20	1.25	70	0.56	120	0.29
25	1.13	75	0.53	125	0.28
30	1.04	80	0.49	130	0.26
35	0.95	85	0.46	135	0.24
40	0.88	90	0.43	140	0.23
45	0.81	95	0.40	145	0.21

Exact determination of the angles g, i, and e is unnecessary for most amateur observational purposes, and sufficient accuracy can be obtained as follows:

let l_s = longitude of sub-solar point (from Sun's selenographic colongitude)

l_e = longitude of sub-Earth point (from librations)

l_p = longitude of area (from map)

then $g \approx l_s - l_e$; $i \approx l_s - l_p$; $e \approx l_p - l_e$ (check: $g = i + e$)

Care should be exercised with the addition and subtraction of these angles, especially when the longitude meridians fall in different quadrant arcs of the equator; negative values for g, i and e should be treated as being positive.

This equation will give the brightness (relative to Full Moon or zero phase angle illumination—but see next paragraph) of a given level area of lunar surface from sunrise to sunset on it. For sloping crater walls or sides of mountains, if the angle of slope is known then angles *i* and *e* can be obtained, but this is a more complicated procedure, requiring the hypothetical translation of the slope to the location at which it would parallel the level lunar surface. This should suffice to indicate possible problems in comparing the brightnesses of craters with their backgrounds, or in trying to relate the brightness of the illuminated portion of a crater's interior to a specific phase angle [as with a crater extinction device (CED)].

The Full Moon Problem

A true Full Moon never occurs from an observing site on Earth, of course, since the Moon is then totally eclipsed. If the integrated light of the Moon, expressed on the stellar magnitude scale, is plotted against phase angle, the portion of the curve for angles lying between 60° and 6° is a straight line; hence extrapolation to phase angle zero is simple and exact, and the tabulated values of k_g given above correspond with this extrapolation. This 'fictitious Full Moon' is the one that has been used until recently for albedo and other determinations; however, it has been known for some time that for phase angles less than about 6°, the lunar surface displays an extra surge of brightness, known as the 'opposition effect', and that it was impossible to extrapolate this accurately. Nevertheless, measurements made on those Apollo photographs which include the anti-solar (i.e., zero phase) point show that the true Full Moon is about 30% brighter (if the measured areas are representative of the whole earthward hemisphere) than the 'fictitious Full Moon'. The same effect is observed on Mercury and the asteroids; the exact cause is not known, but is almost certainly due in part to the "beaded screen" effect of the glass beads in the regolith. This effect, plus the extreme difficulty of determining the absolute luminosities of the Sun and Moon on the stellar magnitude scale, have introduced uncertainties into all determinations of lunar surface albedo, both integrated over the whole disk and also for discrete regions.

Albedo

Albedo (literally, 'whiteness') has several definitions. The original one, proposed by Bond, is the ratio of total solar radiation scattered from a body to the radiation incident upon it. For ordinary visual photometric purposes, it is necessary to modify 'total solar radiation' to 'total visual solar radiation', which gives the *visual Bond albedo* (A_v). This quantity can be determined directly only for bodies that go through a full cycle of phases (Moon, Mercury, and Venus): its value for the Moon is 0.072, but this figure applies only to the earth-ward hemisphere.

This has led to the erroneous, but frequently repeated, statement that the average area of the lunar surface scatters only 7% of the incident light and is therefore darker than all but the blackest of terrestrial rocky materials. The low value is a result of the photometric effects of the porous surface layer, whereby the shadows cast by the particles constitute a substantial percentage of the area viewed. Since no common terrestrial materials have a comparable layer, the Bond albedo value cannot be used for comparison purposes.

Another definition is the *visual geometric albedo* (p_v) which is the ratio of the integrated visual light received from an illuminated body *at zero phase angle* to that from a *plane* white Lambert surface of the same projected area that is normal (at right-angles) to the incident light (at the same distance from the Sun, of course), i.e., $i = e = g = 0°$. For the Moon, the 'fictitious' versus 'true' Full Moon problem presents itself again; the best value for the 'true' Full Moon (i.e., including the opposition effect) is about 0.125, and for the 'fictitious' Full Moon about 0.096. Both because of the impossibility of measuring the integrated light at zero phase angle for Mercury and Venus, and also of knowing whether the few spot measurements of the Moon's opposition effect may be taken as a mean value for the whole disk, the *geometric albedo at 5° phase angle* is now being used more frequently; the value for the lunar nearside would be about 0.084, but this value cannot be used for comparison with terrestrial surfaces for the same reasons that the Bond albedo cannot.

The third and most useful definition of albedo for visual observers is the *visual normal albedo* (ρ_{Ov}) which is the ratio of the brightness of a given area of lunar or planetary surface, or terrestrial sample, at zero phase angle and oriented normal to the incident light, to a plane white Lambert surface similarly illuminated and oriented. For the Full Moon and other rocky atmosphereless bodies, all of which have basically similar photometric properties, the surface need not be oriented at right-angles to the incident light; zero phase angle brightness is unaffected by varying the orientation angle. Thus the Moon's visual geometric albedo (nearside) is simply the mean visual normal albedo of the mix of lighter and darker areas.

The zero phase angle problem affects ρ_{ov} just as much as p_v, of course, and this has introduced the 5° phase angle concept to this parameter also, especially for the planet Mercury. Unfortunately it has been named the 'normal albedo at 5° phase angle', which is a total contradiction of terms!

The following table gives approximate *fictitious* visual normal albedo values for a small selection of lunar features; these values may be compared directly with albedo values of terrestrial materials. For 'true' values (i.e., with opposition effect included), add 0.03.

Darkest areas (W. of Littrow, W. and N. of Bode etc.)	0.083
M. Tranquillitatis (S. of Plinius)	0.088
Plato floor	0.093
M. Serenitatis (E. of Linné)	0.097
M. Imbrium (S. of Plato)	0.101
M. Nectaris, average	0.111
Ptolemaeus floor	0.128
Arzachel, ray-free highlands	0.14
Tycho, general ejecta to NE.	0.17
Aristarchus, mean of interior	0.19
Cassini's "Cloud" in Deslandres	0.21
Proclus, E. wall	0.25
Stevinus A, Abulfeda E etc.	0.27

COLORIMETRIC PROPERTIES

When viewing the Full Moon in a dark sky, it is difficult to believe that it is anything other than an almost silvery white disk with pale grey stains. We have already seen that it is actually comparatively dark in tone, with an albedo averaging slightly under 10% in the visual range, but spectrophotometry shows that it is also not a neutral scatterer; it reflects about 8% of the incident blue light and 12% of the red. This means that it has an overall brownish tint, although different areas of the Moon show small but distinct variations from this. Thus in binoculars or a small telescope with low power, the steely grey colour of M. Tranquillitatis contrasts noticeably with the warmer, buff-coloured tint of M. Serenitatis. Spectrophotometry shows that while the blue normal albedos of these two mare surfaces are about the same (8%), the red normal albedo of the latter is greater by a factor of 1.08 than that of the former (i.e., 12% vs. 11%). The greatest relative blue-red contrasts are about twice as great as this (factor 1.15), but are mostly confined to small isolated areas that are not juxtaposed, so that visual comparison is much more difficult. The reflectivity curve between blue and red is seldom a straight line; some areas show up to a 5% excess in the green region while others show a slight (up to 2%) deficit. Such variations in comparative 'greenness' should be just detectable by observers having well-developed colour discrimination.

Subtractive photographic or other imaging techniques enhance colour differences and eliminate general albedo differences, so that areal variations in colour can easily be seen and related to surface topography. A small photograph showing such a colour difference can be seen in the B.A.A. *Journal* (**90**, 417). The lunar highlands show only very slight colour variations, but the maria display distinct differences among each other and especially among different areas of the same mare. The Mare Imbrium in particular shows sharp, clear-cut boundaries; the greyer areas are found to be younger, titanium-rich lava flows that have overridden older, browner flows. However, of more interest to the observer are the colours of craters and other smaller features, especially when they contrast with their surroundings. The following tables list a number of formations that show such contrasts; they should give permanent 'blinks' when observed with a blink apparatus.

Grey formations on brown background	*Brown formations on grey or intermediate background*
Maraldi floor	Messier and A
J. Caesar, N. floor	Jansen B
Boscovich floor	Moltke
Plato J and ejecta	Plinius
Plato M and ejecta	Lassell G
Pitatus, S. floor	Plato D
Campanus, N. floor	Bullialdus floor
Aristarchus floor and central peak	Carlini
	Darney chi etc. (islands)
	Montes Riphaeus, S. end
	Island with Herigonius eta and pi
	Mons Hansteen alpha

ABOUT THE AUTHOR

While working as a professional astronomer at the Royal Greenwich Observatory, England, Ewen Whitaker developed his interest in the Moon. He has since made many valuable contributions to the study and cartography of the lunar surface. A past Director of the B.A.A. Lunar Section, he took up an appointment at the University of Arizona where he assisted in the planning of landing sites for the Surveyor missions. In 1982 he was awarded the Goodacre Medal of the B.A.A.

RECOMMENDED FURTHER READING

by C.J. Watkis

Baldwin, Ralph B. *The Face of the Moon*. Chicago and London: University of Chicago Press, 1949.

Cameron, Winifred, *Lunar Transient Phenomena Catalog*. Maryland: NASA-Goddard Space Flight Center, 1978.

Charrington, Ernest. *Exploring the Moon Through Binoculars*. London: Peter Davies, 1969.

Fielder, Gilbert. *Lunar Geology*. Guildford, England: Lutterworth Press, 1965.

Fielder, Gilbert, *Structure of the Moon's Surface*. Oxford and New York: Pergamon Press, 1961.

Firsoff, V.A. *The Strange World of the Moon*. London: Hutchinson, 1959.

French, Bevan M. *The Moon Book*. New York: Penguin Books, 1977.

Guest, J.E., and Greeley R. *Geology on the Moon*. London: Wykeham Publications, and New York: Crane-Russak Co., 1977.

Moore, Patrick. *Guide to the Moon*. Guildford, England: Lutterworth Press, 1976.

Rackham, Thomas. *Astronomical Photography at the Telescope*. London: Faber & Faber, 1972.

Taylor, Stuart R. *Lunar Science—A Post Apollo View*. Oxford and New York: Pergamon Press, 1975.

Readers should also consult the *Journal of the British Astronomical Association, Sky & Telescope,* and *Astronomy*.

APPENDIX I:
NUMERICAL FACTS ABOUT THE MOON
by G.W. Amery

Distance from Earth	
maximum	405,547 km (252,007 miles)
minimum	363,263 km (225,732 miles)
mean	384,404 km (238,870 miles)
Sidereal period	27.321661 days
Synodic period	29.530588 days
Axial inclination (referred to ecliptic)	1°32'
Orbital eccentricity	0.0549
Orbital inclination	5°09'
Mean orbital velocity	3684 km/hr (1.02 km/sec) [2289 miles/hr or 0.63 miles/sec]
Apparent diameter	
maximum	33'31"
minimum	29'22"
mean	31'5"
Magnitude of Full Moon (at mean distance)	- 12.7
Mean albedo	0.07
Diameter	3476 km (2,160 miles)
Mass	7.35×10^{25} grams 0.0123 of Earth mass
Volume	0.0203 Earth
Escape velocity (at surface)	2.38 km/sec (1.5 miles/sec)
Density	3.34 gm/c.c. (water = 1)
Surface gravity	0.1653 Earth
Maximum librations	
Geocentric in longitude	7°54'
Geocentric in latitude	6°50'

APPENDIX II:
CHRONOLOGICAL LISTING
OF ALL KNOWN LUNAR PROBES

by E.A. Whitaker

spacecraft		launch date	brief remarks on mission
Thor-Able	1	1958 Aug 17	Failed
Pioneer	1	1958 Oct 11	Failed
Pioneer	2	1958 Nov 8	Failed
Pioneer	3	1958 Dec 6	Failed
Luna	1	1959 Jan 2	Failed
Pioneer	4	1959 Mar 3	Failed
Luna	2	1959 Sep 12	First s/c to reach lunar surface
Luna	3	1959 Oct 4	First s/c to photograph eastern part of farside
Atlas-Able	4	1959 Nov 26	Failed
Atlas-Able	5A	1960 Sep 25	Failed
Atlas-Able	5B	1960 Dec 15	Failed
Ranger	3	1962 Jan 26	Failed
Ranger	4	1962 Apr 23	Failed
Ranger	5	1962 Oct 18	Failed
anon.		1963 Jan 4	Failed
Luna	4	1963 Apr 2	Failed
Ranger	6	1964 Jan 30	Failed
Ranger	7	1964 Jly 28	First s/c to obtain close-up photos (Mare Cognitum)
Ranger	8	1965 Feb 17	Close-up photos of Mare Tranquillitatis (southern part)
Kosmos	60	1965 Mar 12	Failed
Ranger	9	1965 Mar 21	Close-up photos of Alphonsus; bearing strength of soil
Luna	5	1965 May 9	Failed
Luna	6	1965 Jun 8	Failed
Zond	3	1965 Jly 18	Photos of western part of farside
Luna	7	1965 Oct 4	Failed
Luna	8	1965 Dec 3	Failed
Luna	9	1966 Jan 31	First soft landing; close-up panoramas of surface near Cavalerius
Kosmos	III	1966 Mar 1	Failed
Luna	10	1966 Mar 31	Physical measurements from orbit
Surveyor	1	1966 May 30	Soft landing near Flamsteed; extensive photography, physical measurements
Explorer	33	1966 Jly 1	Failed
Lunar Orbiter 1		1966 Aug 10	High resolution stereo photos of selected areas from orbit
Luna	11	1966 Aug 24	Physical measurements from orbit
Surveyor	2	1966 Sep 20	Failed
Luna	12	1966 Oct 22	Physical measurements and photos from orbit
Lunar Orbiter 2		1966 Nov 6	Same as Lunar Orbiter 1
Luna	13	1966 Dec 21	Soft landing near Seleucus; close-up panoramas and physical measurements
Lunar Orbiter 3		1967 Feb 4	Same as Lunar Orbiter 1
Surveyor	3	1967 Apr 17	Soft landing SE of Lansberg; extensive photography, physical measurements
Lunar Orbiter 4		1967 May 4	Medium resolution, large areal coverage photos from orbit

spacecraft		launch date	brief remarks on mission
Surveyor	4	1967 Jly 14	Failed
Explorer	35	1967 Jly 19	Extensive magnetic measurements in Earth-Moon environment
Lunar Orbiter	5	1967 Aug 1	Same as Lunar Orbiter 1, except mostly scientific and farside targets
Surveyor	5	1967 Sep 8	Soft landing E of Sabine; extensive photography, physical measurements, chemical composition
Surveyor	6	1967 Nov 7	Soft landing in Sinus Medii; extensive photography, physical measurements, chemical composition
Surveyor	7	1968 Jan 7	Soft landing on Tycho rim; extensive photography, physical measurements, chemical composition
Zond	4	1968 Mar 2	Unknown objectives
Apollo	6	1968 Apr 4	Failed
Luna	14	1968 Apr 7	Gravity field measurements
Zond	5	1968 Sep 15	First circumlunar flight with return to Earth; physical measurements
Zond	6	1968 Nov 10	Photos of western farside during flyby; films recovered from Indian Ocean
Apollo	8	1968 Dec 21	First manned circumlunar flight; color and b & w photography, visual observations
Apollo	10	1969 May 18	Color and b & w photography, visual observations
Luna	15	1969 Jly 13	Test of automatic navigation system
Apollo	11	1969 Jly 16	First manned landing, E of Sabine; photos from surface and orbit, surface samples, physical measurements
Zond	7	1969 Aug 8	Photos of western farside during flyby; films recovered
Apollo	12	1969 Nov 14	Manned landing at Surveyor 3; photography, surface samples, many physical measurements
Apollo	13	1970 Apr 11	Failed; photography during flyby
Luna	16	1970 Sep 12	Soft landing in E Mare Fecunditatis; sample return
Zond	8	1970 Sep 20	Photos of W farside during flyby; films recovered
Luna	17	1970 Nov 10	Soft-landed Lunokhod 1 (roving vehicle) near Prom. Heraclides; photos and physical data
Apollo	14	1971 Jan 31	Manned landing near Fra Mauro; extensive photos, surface samples and physical data
Apollo	15	1971 Jly 26	Manned landing at Apennines-Hadley site; extensive photos, samples and physical data; first use of Lunar Rover
Luna	18	1971 Sep 2	Physical measurements with automatic navigation system
Luna	19	1971 Sep 28	Same as Luna 18
Luna	20	1972 Feb 14	Sample returned from highlands N of Mare Crisium
Apollo	16	1972 Apr 16	Manned landing at Descartes site; extensive photos, surface samples and physical data
Apollo	17	1972 Dec 7	Manned landing at Taurus-Littrow site; extensive photos, surface samples and physical data
Luna	21	1973 Jan 8	Soft-landed Lunokhod 2 in Le Monnier; photos and physical data
Explorer	49	1973 Jun 10	Radio astronomy from farside (space) of Moon
Luna	22	1974 May 29	Photos and physical data from orbit
Luna	23	1974 Oct 28	Failed; landed in Mare Crisium but no sample return
Luna	24	1976 Aug 9	Sample returned from central Mare Crisium

<u>Landing and impact locations of Lunar Spacecraft and their Modules</u>

s/c		module	long.°		lat.°		landing
Luna	2	/	0		30	N	impact
Luna	5	/	25	W	1.6	S	impact
Luna	7	/	47.8	W	9.8	N	impact
Luna	8	/	62	W	9.6	N	impact
Luna	9	/	64.37	W	7.13	N	soft
Luna	13	/	63.05	W	18.87	N	soft
Luna	15	/	Mare Crisium				impact
Luna	16	DS	56.30	E	0.68	S	soft
Luna	17	DS	35.00	W	38.28	N	soft
Luna	18	/	56.50	E	3.57	N	impact
Luna	20	DS	56.55	E	3.53	N	soft
Luna	21	DS	30.45	E	25.85	N	soft
Luna	23	/	Mare Crisium				soft
Luna	24	DS	62.20	E	12.75	N	soft
Ranger	4	/	129.1	W	12.9	S	impact
Ranger	6	/	21.5	E	9.4	N	impact
Ranger	7	/	20.61	W	10.60	S	impact
Ranger	8	/	24.77	E	2.64	N	impact
Ranger	9	/	2.36	W	12.79	S	impact
Orbiter	1	/	160.71	E	6.35	N	impact
Orbiter	2	/	119.1	E	2.9	N	impact
Orbiter	3	/	91.7	W	14.6	N	impact
Orbiter	5	/	83.1	W	2.8	S	impact
Surveyor	1	/	43.22	W	2.45	S	soft
Surveyor	2	/	11.0	W	4.0	S	impact
Surveyor	3	/	23.34	W	2.97	S	soft
Surveyor	4	/	1.39	W	0.45	N	impact
Surveyor	5	/	23.20	E	1.42	N	soft
Surveyor	6	/	1.40	W	0.53	N	soft
Surveyor	7	/	11.47	W	40.86	S	soft
Apollo	11	DS	23.46	E	0.80	N	soft
Apollo	12	DS	23.42	W	3.04	S	soft
Apollo	12	AS	21.20	W	3.94	S	impact
Apollo	13	S4B	27.86	W	2.75	S	impact
Apollo	14	DS	17.48	W	3.65	S	soft
Apollo	14	AS	19.67	W	3.42	S	impact
Apollo	14	S4B	26.02	W	8.09	S	impact
Apollo	15	DS	3.66	E	26.08	N	soft
Apollo	15	AS	0.25	E	26.36	N	impact
Apollo	15	S4B	11.81	W	1.51	S	impact
Apollo	16	DS	15.51	E	8.97	S	soft
Apollo	16	S4B	23.8	W	1.3	N	impact
Apollo	17	DS	30.77	E	20.17	N	soft
Apollo	17	AS	30.50	E	19.96	N	impact
Apollo	17	S4B	12.31	W	4.21	S	impact

DS = descent stage
AS = ascent stage
S4B = upper rocket stage

<u>Lunar-orbiting s/c that have probably impacted the Moon</u>

Luna 10, 11, 12, 14, 19, 22
Orbiter 4
Apollo 11 AS, Apollo 16 AS

APPENDIX III:
MAP OF THE MOON
by B. Chapman

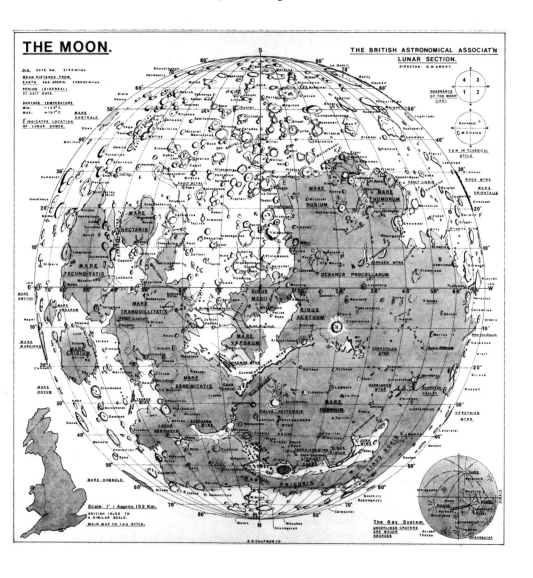

THE MOON.

DIA. 3476 Km. 2160 miles.

**MEAN DISTANCE FROM
EARTH** 384,500 Km. 238000 miles.

PERIOD (SIDEREAL)
27.3217 DAYS.

SURFACE TEMPERATURE
MIN. −153° C
MAX. +101° C

ɣ **INDICATES LOCATION
OF LUNAR DOMES.**

1	

Boussingault 70°
Heimholtz
60°
Biela
Hanno 50°
Pi
MARE
AUSTRALE
Janssen
Vega Rheita Valley Fabricius
Oken Young Metius
Marinus 40° Rheita Riccius
Furnerius Rabbi Levi
Adams Stevinus Neander
30° Snellius Valley Piccolomi
Humboldt Petavius A
Wrottesley Biot
Hecataeus Fracastorius Polybiu
20° Holden MARE Cat
Balmer Vendelinus Beaumont
Cook Rosse
Colombo NECTARIS Cyri
Bellot Madler Theophilus
10° Langrenus Isidorus
Gutenberg Alfraganus
Capella Torricelli
Gilbert MARE
FECUNDITATIS Lubbock
Messier A Censorinus
IAU Webb
E 70° 60° 50° 40° 30° 2
MARE Sabine Ritter
SMYTHII Maskelyne
MARE MARE
Lamont
PYRENAEUS MTNS

THE BRITISH ASTRONOMICAL ASSOCIAT'N

LUNAR SECTION.

DIRECTOR: G.W.AMERY.

3

QUADRANTS
OF THE MOON
(IAU)

4	3
1	2

S / E / W / N

Gassendi
M.Crisium

W / E / S / N

E & W IN CLASSICAL
STYLE

ROOK MTNS

MARE
ORIENTALIE

e Gentil
70°
Bailly
Hausen
60°
Schiller
Phocylides
Wargentin 50°
Nasmyth
Inghirami
ontanus
Mee
Schickard
ari
elm
Drebbel
Lehmann
40°
Clausius
Piazzi
uer
Elger
Capuanus
Lagrange
chus
Ramsden
30° Krasnov
Vitello
Mercator
Vieta
A
Campanus
Leo
Doppelmayer
Cape
LIEBIG FAULT
Cavendish
Byrgius
Kelvin
Hippalus
MARE
Lamarck
önig
Eichstadt
HUMORUM
Mersenius
20°
Agatharchides
Zupus
Darwin
us
Crüger
Gassendi
Darney
A
B
Billy
Sirsalis Rille
Herigonius
Sirsalis
10°
land
Euclides
Letronne
auro
RIPHAEN MTNS
Damoiseau
Grimaldi
Flamsteed
ANUS PROCELLARUM
D
Riccioli
c
IAU
Lansberg
Lohrmann
W
20°
30°
40°
50°
60°
70°
rt
Reinhold
Encke
Hevelius
Suess
Hortensius
Cavalerius
Fauth
Reiner
Olbers
Kepler

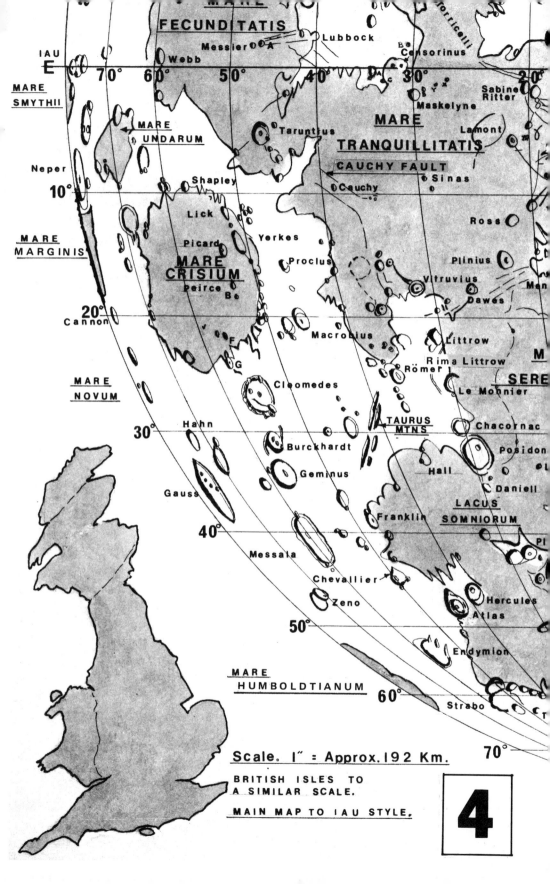

MARE
FECUNDITATIS

Lubbock

Messier
Webb   B•
                                        Censorinus

IAU
E   70°   60°   50°   40°   30°   20°

Maskelyne

MARE
SMYTHII   Sabine
                                        Ritter

MARE   Lamont
UNDARUM   Taruntius

Neper   MARE
                                        TRANQUILLITATIS

10°   Shapley   CAUCHY FAULT
                                        Sinas
                                        Cauchy

Lick   Ross

MARE   Picard   Yerkes   Plinius
MARGINIS
       MARE   Proclus   Vitruvius
       CRISIUM   Dawes
       Peirce   B•   Men

20°   M
Cannon   F   Macrobius
          G   Littrow
                                       Rima Littrow
MARE   Römer I   SERE
NOVUM   Cleomedes   Le Monnier

       Hahn   Chacornac
30°   Burckhardt   Posidon
       Gauss   Geminus   Hall   Daniell

                                       LACUS
                       Franklin   SOMNIORUM
40°   Messala   PI

       Chevallier   Hercules
       Zeno   Atlas
50°
                                       Endymion

       MARE
       HUMBOLDTIANUM   60°
                                       Strabo   T
                                                     70°

Scale. 1″ = Approx. 192 Km.

BRITISH ISLES TO
A SIMILAR SCALE.

MAIN MAP TO IAU STYLE.

4

NUS PROCELLARUM
Flamsteed
D
c Lansberg
Riccioli
IAU
Lohrmann
W
20° 30° 40° 50° 60°

Reinhold
Encke
Hevelius
Suess
Hortensius
Cavalerius
Fauth
Kepler
Reiner
Copernicus
Olbers
10°
Rima Cardanus
Marius
Cardanus
CARPATHIAN
MTNS
Rima Marius
Krafft
Struve
Pytheas
Brayley
20°
Euler
Herodotus
Aristarchus
HARBINGER
MTNS
Prinz
SCHRÖTER'S
VALLEY
Russell
Lambert
La Hire
Krieger
Delisle
Lichtenberg
30°
Heis
HERCYNIAN
Carlini
MTNS
Cape Heraclides
Lavoisier
con
JURA
MTNS
Rümker
6
NUS
DUM
Sharp
Bunsen
Dechen
50°
Tycho
Byrgius A
Lalende
60°
Mösting A
Afraganus
South (1)
Babbage (2)
Kepler
Olbers
Proclus
Copernicus
70°
Pytheas
Timocharis
Carpenter
Autolycus
Lichtenberg
The Ray System.
Timaeus
UNDERLINED CRATERS
ARE MAJOR
SOURCES
Thales
Anaxagoras

ABOUT THE B.A.A.
by C.A. Ronan

The British Astronomical Association was founded in 1890 by a group of amateur and professional astronomers. Their aim was to encourage co-operative work and a general interest in astronomy, and to this end the Association publishes a journal and holds regular meetings in London and elsewhere in Britain. Its membership is not restricted to citizens of the United Kingdom but is open to anyone genuinely interested in astronomy. The Association's headquarters are at Burlington House, Piccadilly, London W1V ONL in that part of Burlington House occupied by the Royal Astronomical Society.

One of the most important ways in which the Association promotes active interest in astronomy is by organizing a series of Observing Sections; these are concerned with the Sun, the Moon, the Terrestrial Planets (Mercury, Venus, and Mars), Jupiter, Saturn, Comets, Meteors, the Aurora, Variable Stars, and Deep-Sky objects. Among the more active of these is the Lunar Section, and it is members of that Section's guiding committee—all of them experienced amateur observers—who have prepared the major part of the material in this publication; the two exceptions are the physicist Dr R.C. Maddison, a former Director of the Section, and Mr Ewen Whitaker who though now a professional lunar astronomer in the United States, was also once Director of the Section.

Membership of the Association, which is open to everyone who is interested in astronomy, is the basic requirement for membership of the Lunar Section. (Forms for membership of the B.A.A. are available from the Association's office at Burlington House.)

INDEX